Innovation Platforms for Agricultural Development

T0091669

Innovation platforms (IPs) form the core of many Agricultural Research for Development programmes, stimulating multi-stakeholder collaboration and action towards the realization of agricultural development outcomes. This book enhances the body of knowledge of IPs by focusing on mature IPs in agricultural systems research, including the crop and livestock sectors, and innovations in farmer cooperatives and agricultural extension services.

Resulting from an international IP case study competition, the examples reported will help the many actors involved with agricultural IPs worldwide reflect on their actions and achievements (or failures), and find tools to share their experience. Chapters feature case studies from Central Africa, Ethiopia, India, Kenya, Nicaragua and Uganda. Authors reflect critically on the impact of IPs and showcase their progress, providing an important sourcebook and inspiration for students, researchers and professionals.

Iddo Dror is Head of Capacity Development at the International Livestock Research Institute (ILRI), Nairobi, Kenya.

Jean-Joseph Cadilhon is Senior Agricultural Economist at ILRI, Nairobi, Kenya.

Marc Schut is a Social Scientist at the International Institute of Tropical Agriculture (IITA), Burundi, and Wageningen University and Research Centre, the Netherlands.

Michael Misiko is an Innovation Scientist at the International Maize and Wheat Improvement Center (CIMMYT), Kenya.

Shreya Maheshwari is an independent editor based in India.

Innovation Platforms for Agricultural Development

Evaluating the mature innovation platforms landscape

Edited by
Iddo Dror, Jean-Joseph Cadilhon,
Marc Schut, Michael Misiko and
Shreya Maheshwari

First published 2016
by Routledge
2 Park Square, Milton Park, Abingdon, Oxon OX14 4RN

and by Routledge
711 Third Avenue, New York, NY 10017

First issued in paperback 2018

Routledge is an imprint of the Taylor & Francis Group, an informa business

© 2016 International Livestock Research Institute

All rights reserved. No part of this book may be reprinted or reproduced or utilised in any form or by any electronic, mechanical, or other means, now known or hereafter invented, including photocopying and recording, or in any information storage or retrieval system, without permission in writing from the publishers.

Trademark notice: Product or corporate names may be trademarks or registered trademarks, and are used only for identification and explanation without intent to infringe.

British Library Cataloguing-in-Publication Data
A catalogue record for this book is available from the British Library

Library of Congress Cataloging in Publication Data
Names: Dror, Iddo, editor.
Title: Innovation platforms for agricultural development/edited by Iddo Dror, Jean-Joseph Cadilhon, Marc Schut, Michael Misiko and Shreya Maheshwari.
Description: New York: Routledge, 2016. | Includes bibliographical references and index.
Identifiers: LCCN 2015032217 | ISBN 9781138181717 (hbk) | ISBN 9781315646817 (ebk)
Subjects: LCSH: Agriculture—Economic aspects—Developing countries. | Economic development—Technological innovatians—Developing countries.
Classification: LCC HD1417 .I556 2016 | DDC 338.109172/4—dc23
LC record available at http://lccn.loc.gov/2015032217

Typeset in Bembo
by Florence Production Limited, Stoodleigh, Devon, UK

ISBN 13: 978-1-138-58890-5 (pbk)
ISBN 13: 978-1-138-18171-7 (hbk)

Contents

Figures

Tables

Boxes

Contributors

Editors

Jean-Joseph Cadilhon, is a Senior Agricultural Economist in the Policy, Trade and Value Chains Program of the International Livestock Research Institute (ILRI). Based in Nairobi, he elaborates methods and tools for value chain analysis and multi-stakeholder platform performance evaluation. He then field-tests them within agri-food development projects in African, Asian and Latin American countries. Jo holds two MSc degrees from AgroParisTech, one in geography and agricultural development, the second in rural public administration. He also holds a PhD in food marketing from Imperial College London and a BSc in Chinese language and civilization from INALCO in Paris. He has been documenting and researching the role of multi-stakeholder platforms in agri-food value chain development for the past seven years when previously working at the United Nations Food and Agriculture Organization Regional Office for Asia and the Pacific, and in the think-tank of the French Ministry of Agriculture. Jo has also co-authored the Nicaragua Learning Alliance case study featuring in this book.

Iddo Dror, is Head of Capacity Development for the International Livestock Research Institute (ILRI, based in Nairobi, Kenya). Iddo leads the development of knowledge, attitudes, skills and institutional arrangements that are necessary to replicate, expand, adapt, use, support and sustain research and its application for development in various contexts across ILRI and its projects. Iddo has more than 15 years' capacity development experience, gained in academia and international development, combined with experience in teaching, research and fellowship administration. His career track has included assignments with UN agencies and the private sector, several years at the University of Geneva setting up and running an innovative MBA programme, and work in India with the Micro Insurance Academy (MIA), an organization he co-founded and helped transition to become an international award-winning, ISO-certified leader in its field, with research and capacity development activities across several Asian and African countries. He has a PhD (Economic and Social Sciences) and MBA from the University of Geneva.

Shreya Maheshwari is a freelance consultant in India and a communications consultant working with international development organizations. She co-authored the first business school case study published by the International Livestock Research Institute (ILRI): the Case Study on Index Based Livestock Insurance catalogues, the development of the world's first insurance for African pastoralists, and has been taught at University of Geneva's International Organizations MBA Program and at the University of Cape Town. Besides ILRI, Shreya has worked with the Clinton Foundation, the International Bank for Reconstruction and Development, the Government of India, ActionAid and several other organizations in the field of economic development. She has been published in *Dispatches from the Developing World* (anthology), *Harvard Political Review* and *The Times of India*. Shreya holds a BA in Economics from Harvard University and a Business Leadership Program Certificate from Harvard Business School.

Michael Misiko is an Innovation Scientist at the International Maize and Wheat Improvement Center (CIMMYT), Kenya. His work includes integrating local skills and scientific research through interactive learning for systems intensification, Agricultural Innovation Systems (AIS), and fostering partnerships for research and scaling. This scaling incorporates private sector agribusiness, information and communication technologies, and appropriate farmer participation. Michael has developed and translated farmer videos for use across Africa. He is an Agricultural Anthropologist with a Masters (University of Nairobi) and a PhD (Wageningen University). He started his career at the Tropical Soil Biology and Fertility Program of the United Nations Educational, Scientific and Cultural Organization. He has been working for several CGIAR centres in Africa since 2002 with a short stint as a head of department at National Museums of Kenya before joining CIMMYT in 2012. He has designed and led several international projects, in-depth studies, research-and-development partnerships with international and national programmes, high-level policy dialogues, and mentored international capacity for AIS.

Marc Schut, a Dutch national, is a Social Scientist working with the International Institute of Tropical Agriculture (IITA) and Wageningen University and Research Centre (WUR), based in Bujumbura, Burundi. Marc holds an MSc in Agro-ecological Knowledge and Social Change and a PhD in Communication and Innovation Studies from Wageningen University. His PhD focused on the role of science in policy and innovation processes. He has supported several IPs. Currently, he coordinates research for development activities on multi-stakeholder processes, scaling and institutional innovation under the CGIAR Research Program on Integrated Systems for the Humid Tropics (Humidtropics). Together with MSc and PhD researchers and development partners, Marc seeks to support and study multi-stakeholder processes in East and Central Africa, and share lessons learned with colleagues worldwide. Marc's ambition is to conduct research that will strengthen agricultural education and innovation systems in developing countries to foster capacity to innovate by (young) farmers and other stakeholders in the agricultural sector.

Authors

Victor Afari-Sefa, World Vegetable Center, Tanzania, co-author of the case study on crop–livestock–tree interaction in Mukono–Wakiso (Uganda) and a citizen of Ghana, is an Agricultural Economist. He is the Global Theme Leader – Consumption at The World Vegetable Center (AVRDC). Operating from AVRDC's Regional Office for Eastern and Southern Africa (Arusha, Tanzania), he is responsible for monitoring the quality of science, and takes the lead in identifying priorities, opportunities and constraints to AVRDC's consumption research and development theme. Victor also leads and coordinates vegetable socio-economic research in sub-Saharan Africa and globally. He assesses opportunities and challenges in production systems, and analyses constraints in value chains and policy in an interdisciplinary context. Victor is the global focal point of AVRDC for the Humidtropics programme.

Christian Andres is the lead author of the case study entitled 'Sustaining the Supply of Organic White Gold'. He is a Swiss national and works as a Researcher in Tropical Production Systems at the Research Institute of Organic Agriculture (FiBL), Switzerland. Christian lived through the innovation process from using the insights of first trials and errors to the achievement of outputs and finally scaling to reach impact (years 7–10 of the IP). This motivated him to spread the good news from the field in this compendium.

Gurbir Singh Bhullar is a co-author of the case study entitled 'Sustaining the Supply of Organic White Gold'. He is an Indian national and works as a Senior Scientist at the Research Institute of Organic Agriculture (FiBL), Switzerland. Gurbir leads FiBL's programme on 'Long-term Farming Systems Comparison in the Tropics' (SysCom), and acts as the main coordinator of the SysCom site in India. He thinks that this case study could serve as a motivation for other IP practitioners, as it is an excellent example of how an IP has helped farmers increase their productivity while also attracting the interest of conventional farmers for organic farming.

Guy Blomme is a Bioversity International, Ethiopia scientist working on germplasm and integrated disease management for more resilient and productive banana and enset-based cropping systems in East and Central Africa, based in Addis Ababa. In the framework of Humidtropics Guy works on understanding farmer-led crop diversification strategies in zones affected by biotic stresses. He also contributes to system intensification endeavours for more resilient and nutritious cropping systems. He co-authored the CIALCA case.

Thomas Dubois (Belgium) is Director of The World Vegetable Center (AVRDC) Eastern and Southern Africa, based in Tanzania. Dr Dubois received his PhD in Entomology from Cornell University (New York) in 2003. He has experience in aflatoxin mitigation activities, tissue culture development for banana marketing systems in East Africa, biopesticides and biotechnologies. His recent activities have emphasized value chains for rice in 11 sub-Saharan African countries, developing banana as an income-generating crop through linkage with the private sector, and enhancing technologies to reduce aflatoxin contamination of crops on a commercial scale. Dr Dubois received the 2006 Young Promising Scientist Award from the CGIAR. He has co-authored the Mukono–Wakiso case study from Uganda in this book.

Alan Duncan works for the International Livestock Research Institute (ILRI) in Ethiopia and is originally from the UK. He is a research scientist with expertise in livestock feeding but is also interested in the connections between scientific knowledge and practical application, and in particular the human angle to those connections. Alan is a contributor to the case study from Ethiopia and has also co-authored the Indian MilkIT case. Alan drove the establishment of both these projects, ensuring that IPs were central to the project plans. The core role of IPs in the projects arose out of a realization that technical change in smallholder systems required a strong emphasis on dialogue and joint action among many different players.

Edison Hilman holds a BSc (Agriculture) from Makerere University (Uganda), and is completing an MSc (Crop Science) there. He has been Agricultural Officer in Bubaare sub-county, Kabale District, Uganda since 1999 and District Agricultural Officer since 2011. Edison is responsible for mobilizing and organizing farmers and communities to identify, control and monitor crop pests and diseases outbreaks. He has experience with various agricultural development approaches: farmer field schools, value chain development and multi-stakeholder platforms. He has facilitated activities at the Bubaare IP featured in this book since 2009. His ambition is to contribute to improving the living standards of communities by promoting innovation and technology in agricultural development, upholding virtues of hard work, devoted service, commitment, patriotism and inspirational leadership.

Celister Kaleha is an administrative officer at Resource Projects Kenya, a non-profit organization that is also a member of the WeRATE platform. She is the Monitoring and Evaluation Specialist for N2Africa in addition to holding the position of secretary to the WeRATE NGO. She promotes soil health technologies and practices among farmers in Western Kenya. She is a co-author of the case entitled 'Humidtropics IP Case Study: WeRATE Operations in West Kenya'.

Rebecca Mutebi Kalibwani, senior lecturer at Bishop Stuart University (Mbarara, Uganda) in the Department of Agriculture and Agribusiness, is lead author of the Bubaare IP case study (Uganda). She is an agricultural economist by profession, with a BSc in Agriculture and an MSc in Agricultural Economics, both from Makerere University. Since 2009 she has facilitated activities on policies and institutional arrangements on the four IPs established by the Sub-Saharan Africa Challenge Program and funded by the Forum for Agricultural Research in Africa (FARA) in Southwestern Uganda. She is particularly interested in following up the institutional development of the IPs and related policy issues.

Rick Kamugisha works with the World Agroforestry Centre (ICRAF) Uganda office and holds a Bachelor's degree (Sociology) and a Master's degree (Development Studies). Rick has been involved in facilitating 'Landcare' and natural resource management, and in strengthening local-level policies in more than 15 districts of Uganda. Rick has worked in past projects proving the concept of International Agricultural Research for Development (IAR4D) in the Lake Kivu region and using the 'Landcare' approach to foster collective action and learning for widespread impact on sustainable land management. He is currently involved in the Humidtropics programme in Uganda and has co-authored the Bubaare IP case study. Rick's ambition is to pursue his further studies in Agricultural Extension and Rural Innovations.

Dieuwke Lamers is co-author of the Ugandan case study 'Mukono–Wakiso'. Currently Dieuwke works for the International Institute of Tropical Agriculture (IITA) in Burundi as a Junior specialist on research for development and IPs. Before this, Dieuwke did a four-month internship at IITA Kampala during which she was highly involved with the Humidtropics platforms in Uganda, including the Mukono–Wakiso IP. She attended platform meetings and activities and interviewed platform members and facilitators about the process. In parallel, she worked a lot on documentation of the platform processes.

Dirk Hauke Landmann, International Livestock Research Institute (ILRI), Kenya, is the first author of the case study entitled 'With Trust and a Little Help from Our Friends: How the Nicaragua Learning Alliance (NLA) Scaled up Training in Agribusiness'. Dirk recently finished his Master of Science at the Georg-August-University of Göttingen with a specialization in Agribusiness. This case study was developed from his research at the ILRI. Dirk is now a PhD candidate in the Research and Training Group 'Global Food' of Göttingen University and analysing business relationships in developing countries. Dirk felt it was important to present insights and lessons learned from his study to help improve the NLA and other similar platforms. The only way to increase knowledge is to share it.

Ewen Le Borgne, International Livestock Research Institute, Ethiopia, is a contributor to the case study from Ethiopia entitled 'Innovation Platforms for Improved Natural Resource Management and Sustainable Intensification'. Ewen is team leader of Knowledge, Engagement and Collaboration in the communication and knowledge management unit of the International Livestock Research Institute (ILRI) in Addis Ababa. At ILRI and with his previous employer, he has worked on the establishment, facilitation, communication process, and monitoring and evaluation of multi-stakeholder processes. Ewen's interest with IPs and other multi-stakeholder processes revolves around the rich social learning that happens between the actors involved and how this leads (under certain conditions) to transformation and real social change in the longer run.

Zelalem Lema, International Livestock Research Institute, Ethiopia, is lead author of the case study from Ethiopia entitled 'Innovation Platforms for Improved Natural Resource Management and Sustainable Intensification'. Zelalem is a Research Officer – Innovation System in Agriculture, working for the International Livestock Research Institute (ILRI) in Addis Ababa. He supports multi-stakeholder processes for different ILRI projects and programmes: engaging various actors at different levels through IPs to bring impact to smallholder farmers. He was involved in initiating and facilitating the platforms of this case study during the Nile Basin Development Challenge Program and supported their nesting into Humidtropics. Zelalem wants to share the experience of Ethiopian IPs in addressing the complex constraints of mixed crop–livestock farming systems while enhancing natural resources.

Lokendra Singh Mandloi has done the bulk of the work on the ground for the Indian case study entitled 'Sustaining the Supply of Organic White Gold'. He is an Indian national and works as a Research Coordinator at the bioRe Association (India). Lokendra has been working very closely with the stakeholders throughout the whole process from problem identification to scaling (years 3–10 of the IP). He taught the farmers how to carry out research on their own farms and supervised them. Last but not least, Lokendra was the main facilitator of the whole process.

Mariëtte McCampbell, an MSc Development and Rural Innovation student from Wageningen University, the Netherlands, is co-author of the Mukono–Wakiso case study. A research project during her Masters introduced her to IPs. Currently Mariëtte works on her thesis in the Policy Action for Sustainable Intensification of Ugandan Cropping systems (PASIC) programme of the International Institute of Tropical Agriculture (IITA), writing about fertilizer adoption constraints in South West Uganda. Previously she did her internship for Q Energy Consultants and IITA, looking at the utilization of bio-digestion systems, while also attending Humidtropics meetings and assisting in a vegetables development project.

Perez Muchunguzi, International Institute of Tropical Agriculture, Uganda, is the lead author for the CIALCA case study on Central and Eastern Africa. He is a Multi-stakeholder Specialist at the International Institute of Tropical Agriculture (IITA) in Kampala (Uganda) where he supports multi-stakeholder processes and partnerships in the Humidtropics programme. He collaborated with his co-authors to write the case study, highly motivated by the fact that this presents an opportunity to share the CIALCA story with a global audience, highlighting both the positive outcomes as well as the remaining challenges: both good learning points for Humidtropics as well as other initiatives. Perez has also co-authored the Ugandan case study 'Mukono–Wakiso'.

Immaculate Mugisa is a co-author of the Mukono–Wakiso case study from Uganda. She is a Crop Agronomist with the National Agricultural Research Organization in Uganda, based at Mukono Zonal Agricultural Research and Development Institute. Her key role is to study the effects of various cropping systems on crop yield and determine the most efficient for optimum productivity. She has wide experience in facilitating multi-stakeholder processes. She was part of the national team that implemented the 'Participatory Market Chain Approach' in Uganda in collaboration with other local, regional and international development partners. Currently, she is one of the national facilitators under the Humidtropics programme's Uganda Action Site. She wishes to share her experience through this case study.

Welissa Mulei is the data manager for WeRATE, Kenya and co-author of 'Humidtropics Innovation Platform Case Study: WeRATE Operations in West Kenya'. She provides administrative and data support services for the research and development partners involved in WeRATE projects to improve beans and cassava, and to eliminate Striga and other pests. Welissa and her co-authors decided to write this case study to document the operations of the WeRATE platform, its impacts and outcomes, successes and failures so as to facilitate wider understanding of the platform, and more direct engagement with it.

Annet Abenakyo Mulema contributed to the case study from Ethiopia entitled 'Innovation Platforms for Improved Natural Resource Management and Sustainable Intensification'. Annet is a sociologist, currently working for the International Livestock Research Institute (ILRI) as a Social Scientist – Gender in Addis Ababa (Ethiopia). Her work focuses on gender research and integration in agricultural programmes related to livestock and intensification of mixed farming systems. Annet is a Ugandan with expertise in gender analysis, value chains, innovation systems, participatory research approaches and food systems. Annet conducted the impact study to assess stakeholders' change in knowledge, attitude, skills and practices as a result of the project interventions described in the case. She is inspired by IPs' potential to transform smallholder farmers' livelihoods.

Sylvia Namazzi is the lead author of the study entitled 'Crop–Livestock–Tree Integration in Uganda: The Case of Mukono–Wakiso Innovation Platform'. Sylvia is a research associate at the World Vegetable Center (AVRDC), based in Uganda at the International Institute for Tropical Agriculture (IITA). She is coordinating AVRDC's activities on Humidtropics in Uganda and is actively involved in platform processes in the Humidtropics field sites of the country. As a team, Sylvia and her co-authors thought it important to share lessons learned on the platform through writing this case study.

Sospeter O. Nyamwaro, Center for Tropical Agriculture, Uganda, is an agricultural economist with PhD and MSc degrees from the University of New England (Australia) and Texas A&M University, respectively. As an employee of the International Center for Tropical Agriculture (CIAT), he has coordinated research for development (R4D) activities in the Lake Kivu Pilot Learning Site of the Sub-Saharan Africa Challenge Program and Humidtropics. He is a co-author of the case study on the Bubaare IP Cooperative Society Limited, one of 12 IPs emerging from this continental R4D initiative. He has also worked as Senior Principal Research Scientist for the Kenya Agricultural and Livestock Research Organization (KALRO), coordinating the National Range and Livestock Research.

Thanammal Ravichandran, International Livestock Research Institute, India, is lead author for the case study from India entitled 'Changing Women's Lives – One Cow and One Litre of Milk at a Time – Deep in the Foothills of India's Himalayan Mountains'. After working for a decade as a veterinarian on the welfare of working horses and donkeys in India, she pursued higher studies in development economics. She joined the International Livestock Research Institute (ILRI) in 2012 for the MilkIT project described in the case. She was based in Uttarakhand, coordinating the selection of sites and partners, setting the IPs, implementing interventions and capturing changes through impact assessment. Thanammal has now started PhD studies with ILRI to study the gender and social dimensions of dairy institutions in India.

Murat Sartas, International Institute of Tropical Agriculture, Uganda, Wageningen University, the Netherlands, and Swedish University of Agricultural Sciences, Sweden, is a scientist specializing in measuring the performance of agricultural innovation systems and agricultural value chains in low and middle-income countries. He holds two BSc degrees in Economics and International Relations from Middle East Technical University (Turkey) and four MSc degrees in Quantitative Economics (Turkey), Agricultural Economics (Sweden), Agribusiness Management (Germany) and Rural Development and Natural Resource Management (Sweden). Since April 2014, he has been doing a PhD at Wageningen University on the effectiveness of multi-stakeholder processes in achieving development outcomes as part of the Humidtropics programme. Murat has completed different research-for-development and monitoring-and-evaluation tasks in projects focused on agricultural-based livelihoods in several African, Asian and European countries. He has co-authored the Ugandan Mukono–Wakiso case study.

Anna Sole-Amat is a co-author of the Ugandan case study on the crop–livestock–tree integration in Mukono–Wakiso Humidtropics site, Uganda. Anna is a biologist, working in science and communication at the International Institute of Tropical Agriculture (IITA). Based in Kampala (Uganda), she has helped facilitate the implementation of Humidtropics at field level since the beginning and to capture the process through monitoring and evaluation tools. Anna and her co-authors thought it very important to share the experiences from the Mukono–Wakiso IP, working within a systems approach rather than on specific commodities.

Moses M. Tenywa, Makerere University, Uganda, holds a PhD in soil and water management from Ohio State University and is a Professor at Makerere University (Kampala, Uganda). Moses was a Principle Investigator and Partner in the Sub-Saharan Africa Challenge Program to achieve proof of the international-agricultural-research-for-development (IAR4D) concept. As part of this regional IAR4D programme, he facilitated the formation of 12 IPs in Uganda, Rwanda and the Democratic Republic of Congo. He is the facilitator of the Ugandan Mukono–Wakiso IP featured in this compendium. He also co-authored the case study featuring the Bubaare IP, also in Uganda.

Nils Teufel works for the International Livestock Research Institute (ILRI) in Kenya; he was based in the Delhi office from 2006 to 2013. He is a German agricultural economist with a special interest in livestock farming systems. Nils has studied dairy-intensifying smallholders in South Asia for several years and aims to highlight the changes that improved access to markets can lead to within these poor households. He co-authored the MilkIT case study as he oversaw the Indian component of this project in the Himalayan hills. MilkIT IPs were interesting as they showed how various interventions into milk marketing led to organizational changes and induced farmers to explore and implement innovative ideas to increase their dairy animals' productivity.

Honest Tumuheirwe, Kabale District Local Government, Uganda, holds a Bachelor of Science degree in Botany and Zoology from Makerere University (Uganda), and a postgraduate Diploma in Project Planning and Management from Kabale University (Uganda). She has been a Fisheries Officer for the Kabale District Local Government since 2000. She facilitates farmer field schools in a project funded by the Food and Agriculture Organization of the United Nations in Bubaare sub-county. She contributed to the Bubaare IP case study featured in this book, thanks to her experience of facilitating agricultural extension activities at the Bubaare IP since 2010.

Jeniffer Twebaze is Acting District Production Officer, Kabale District Local Government (Uganda). She oversees the sectors of agriculture, livestock, fisheries and commerce. She holds a postgraduate diploma in Project Planning and Management, Bachelor of Science in Botany and Zoology and she is currently pursuing a Master degree in Zoology at Makerere University (Uganda). Jeniffer has been interacting with rural farmers since 1999 when she started working with the Kabale District Local Government at the lowest level of a sub-county. She has been the District focal point person for the Sub-Saharan Africa Challenge Program in which the Kabale Local Government was participating. Jeniffer has been facilitating all Bubaare IP activities since 2009 and co-authored the platform's case study.

Piet van Asten is a systems agronomist at the International Institute of Tropical Agriculture (IITA) in Uganda. He has been working on sustainable intensification of perennial-based cropping systems in Africa's humid zones for the past ten years. He has over 17 years of professional research experience in sub-Saharan Africa and holds a PhD in soil science and agronomy from Wageningen University (the Netherlands). He is increasingly involved in managing and supporting research-for-development projects on a regional scale that embed multi-stakeholder platforms tackling various issues from the soil pit to household economics, linkages to input–output markets, and climate change mitigation and adaptation. He is a co-author of the two case studies on CIALCA and Mukono–Wakiso IPs.

Bernard Vanlauwe joined the International Institute of Tropical Agriculture (IITA) in Kenya in 2012 to lead the Central Africa hub and the Natural Resource Management research area. In this capacity, he also oversees activities for Humidtropics and other CGIAR research programmes. He has previously worked for various international and national research institutions with a focus on soil and environmental sciences in sub-Saharan Africa. He holds a PhD in Applied Biological Sciences from the Catholic University of Leuven (Belgium). He participated in the CIALCA platform featured in this compilation while working for the Tropical Soil Biology and Fertility research area of the International Center for Tropical Agriculture (CIAT).

Paul L. Woomer, International Institute of Tropical Agriculture, Kenya, holds a PhD in Agronomy and Soil Science. He is the first author of the 'Humidtropics Innovation Platform Case Study: WeRATE Operations in West Kenya'. He is a US citizen and 25-year Kenyan resident, currently working with the International Institute of Tropical Agriculture (IITA), Kenya. He is a Project Scientist, N2Africa Kenya Country Coordinator and Humidtropics Action Research Scientist. His fields of expertise are biological nitrogen fixation and legume agronomy. Paul serves as a part-time Technical Advisor to the WeRATE Platform and, as one of the principle architects of

Humidtropics, Paul seeks to understand better the potential of research-for-development platforms as key linkages between on-farm technology testing and larger-scale collective action and institutional innovation.

Foreword

The field of Agricultural Research for Development (AR4D), in which CGIAR is an important player, is continuously exploring strategies to further increase their development impacts at scale. In the CGIAR Strategy and Results Framework (SRF) 2016–2030 the consortium's mission has been defined as:

> to advance agri-food science and innovation to enable poor people, especially poor women, to increase agricultural productivity and resilience; share in economic growth and feed themselves and their families better; and manage natural resources in the face of climate change and other threats.

Research drives innovation to generate new and improved technologies (e.g. better seeds, machinery or management practices) as well as institutions (e.g. policies, new modes of collaboration). Both can enhance capacity to innovate or create an enabling environment for people to identify, prioritize and solve their own problems.

New technology does not automatically lead to impact at scale. Users only accept and adopt new technology if the new solution responds to their demand. The new research for development paradigm holds that simply producing new knowledge and making it available as global public goods is not good enough. The likelihood of achieving impact at scale improves if users have been involved in research from its conceptualization, and if research organizations develop strategic partnerships to ensure that the knowledge generated by research can move down the impact pathway, lead to innovation, lead to products in the market place, and lead to uptake and use. One mechanism to foster involvement of all stakeholders in the agri-food value chain, end-users, government and the private sector, is an IP approach (also sometimes referred to as learning alliance, multi-stakeholder platform etc.). Such platforms and partnerships are essential to foster research for development efforts towards innovations that lead to impact at scale.

Several AR4D programmes, including the CGIAR Research Program on Integrated Systems for the Humid Tropics (Humidtropics) and other CGIAR research programmes, have been promoting IPs to increase the efficiency and impact of their activities in order to move from research outputs to development outcomes. It is expected that IPs and other multi-stakeholder approaches will continue to play an important role in the second phase of CGIAR research programmes that will kick off in 2017.

This book assesses and reflects on the performance of mature IPs. It aims to promote learning and sharing through experiences with these platforms. We think it will play an important role in linking the theory and practice of IPs. The case studies from three continents provide insight in how facilitation and other platform support functions can effectively link multi-stakeholder processes to innovation that leads to durable development outcomes in fields ranging from productivity improvement and natural resources management to institutional innovation. In doing so, we believe the authors have succeeded in producing a book that will advance the debate on what type of investments, competences and partnership strategies are needed to help produce healthy diets from sustainable agri-food systems for all.

Frank Rijsberman
CEO CGIAR Consortium

Kwesi Atta-Krah
Director Humidtropics

Acknowledgements

This book was developed under the CGIAR Research Program on Integrated Systems of the Humid Tropics (Humidtropics).

We would like to thank:

The CGIAR Research Program on Integrated Systems for the Humid Tropics (Humidtropics), the International Livestock Research Institute (ILRI), the Forum for Agricultural Research in Africa (FARA), and the International Institute of Tropical Agriculture (IITA) for funding this work.

All case study authors for sharing their enthusiasm and experiences in this book.

Susan MacMillan, Sara Quinn and Valerie Poire for their fantastic support on communications, and for encouraging the case study authors to develop their thoughts and storyline, as well as the creative support of Brooza Studio who created the case study illustrations.

Jennifer Kinuthia for wonderful administrative support throughout the project – and for her endless patience with all editors and authors.

Abbreviations

AAS	Agricultural Aquatic Systems
AATF	African Agricultural Technology Foundation
ACE	Area Cooperative Enterprise
AdA	Learning Alliance – Alianza de Aprendizaje
AG	Aktiengesellschaft (German term for 'incorporated company')
AGINSBA	Agriculture Innovation System Brokerage Association
AGRA	Alliance for a Green Revolution in Africa
AH	animal husbandry
AI	artificial insemination
AIS	Agricultural Innovation System
ANOVA	analysis of variance
AR4D	Agricultural Research for Development
ARDAP	Appropriate Rural Development Agriculture Program
ASAE	Asian Society of Agricultural Economists
BAIF	Bharatiya Agro Industries Foundation
BCIE	Central American Bank for Economic Integration – Banco Centroamericano de Integración Económica
BM	buttermilk
BNF	Biological Nitrogen Fixation
BoT	board of trustees
BTC	Belgium Technical Cooperation
BUFFSO	Butula Farmers' Field School Organization
BUSCO	Butere Soybean Cooperative
BUSSFFO	Bungoma Small Scale Farmers
CATIE	Tropical Agricultural Research and Higher Education Center – Centro Agronómico Tropical de Investigación y Enseñanza
CBO	community-based organizations
CCAFS	Climate Change, Agriculture and Food Security
CEO	Chief Executive Officer
CGIAR	previously known as the Consultative Group on International Agricultural Research

CIALCA	Consortium for Improvement of Agricultural-based Livelihoods in Central Africa
CIAT	International Center for Tropical Agriculture
CIMMYT	International Maize and Wheat Improvement Centre
CIP	International Potato Center
CPWF	Challenge Programme for Water and Food
CRS	Catholic Relief Service
CSR	Cooperative Society Regulations
DC	District Commercial Officer
DCO	District Commercial Office
DDC	Belgium Development Cooperation
DRC	Democratic Republic of the Congo
ETB	Ethiopian Birr
FAO	Food and Agriculture Organization of the United Nations
FARA	Forum for Agricultural Research in Africa
FEAST	Feed Assessment Tool
FENACOOP	National Federation of Agricultural Cooperatives and Agribusiness – Federación Nacional de Cooperativas Agropecuarias y Agroindustriales
FiBL	Forschungsinstitut für Biologischen Landbau (German term for Research Institute of Organic Agriculture)
FORMAT	Forum for Organic Resource Management and Agricultural Technologies
FUNICA	Foundation for Technological Development of Agriculture and Forestry of Nicaragua – Fundación para el Desarrollo Tecnológico Agropecuario y Forestal de Nicaragua
FYM	farmyard manure
GDP	gross domestic product
GIZ	German Federal Enterprise for International Cooperation – Deutsche Gesellschaft für Internationale Zusammenarbeit
HAGONGLO	Health Agriculture General Orientation Nutrition Generation Livelihood Overcomers
HECOP	Heritage Conservation Promotion Organization
IAR4D	Integrated Agricultural Research for Development
IAR-IN-D	International Agricultural Research-IN-Development
ICRAF	World Agroforestry Center
ICT	information and communication technology
IDO	intermediary development outcomes
IDRC	International Development Research Center
IFAD	International Fund for Agricultural Development
IFPRI	International Food Policy Research Institute
IITA	International Institute for Tropical Agriculture
ILRI	International Livestock Research Institute
INR	Indian Rupees

INTA	Nicaraguan Institute of Agricultural Technology – Instituto Nicaragüense de Tecnología Agropecuaria
IP	innovation platform
IR maize	imazapyr-resistant maize
ISFM	integrated soil fertility management
IWMI	International Water Management Institute
KALRO	Kenya Agricultural and Livestock Research Organization
KARI	Kenya Agricultural Research Institute
KASP	knowledge, attitudes, skills and practices
KAZARDI	Kabale Zonal Agricultural Research and Development Institute
KDLG	Kabale District Local Government
KENAFF	Kenya National Farmers Association
KESOFA	Kenya Soybean Farmers Association
KHG	Kleen Homes and Gardens
KUFGRO	Kuria Farmers Group
KVK	Krishi Vikas Kendra
LA	Learning Alliance
LED	Liechtenstein Development Service
LKPLS	Lake Kivu Pilot Learning Site
LTE	long-term farming systems comparison experiment
LV	local vinegar
LWR	Lutheran World Relief
MAGFOR	Ministry of Agriculture and Forestry – Ministerio de Agricultura y Forestal
MDG	Muungano Development Gateways
MEFCCA	Ministry of Family, Community, Cooperatives and Associative Economics
MFAGRO	Mwangaza Farmers Group
MFI	Micro Finance Institution
MIA	Micro Insurance Academy
MINAGRI	Ministry of Agriculture
MLNV	Maize Lethal Necrosis Virus
MSc	Master of Science
MTTI	Ministry of Trade, Tourism and Industry
MUDIFESOF	Mumias District Federation of Soybean Farmers
NABARD	National Bank for Agriculture and Rural Development
NARO	National Agriculture Research Organization
NARS	National Agricultural Research Systems
NBDC	Nile Basin Development Challenge
NGO	non-governmental organization
NIFA	National Institute of Food and Agriculture
NLA	Nicaragua Learning Alliance
NOGAMU	National Organic Agriculture Movement in Uganda
NRM	natural resource management

OLM	outcome logic model
OWDF	One World Development Foundation
PAR	participatory action research
PhD	Doctor of Philosophy
POR	participatory on-farm research
R4D	research for development
RAB	Rwanda Agricultural Board
ROP	Rural Outreach Program
RP	rock phosphate
RP–FYM	rock phosphate enriched FYM
RPK	Resource Projects Kenya
RPO	Rural Producer Organization
SACCO	Savings and Credit Cooperative
SCC-VI	Swedish Cooperative Centre
SCODP	Sustainable Community Development Program
SCP	Structure–Conduct–Performance
SDC	Swiss Agency for Development and Cooperation
SHG	Self Help Group
Sig.	significance
SNV	Foundation of Netherlands Volunteers – Stichting Nederlandse Vrijwilligers
SRF	strategy and results framework
SSA CP	Sub-Saharan Africa Challenge Program
SWC	soil and water conservation
ToT	training of trainers
UCA	Union of Jinotega Agricultural Cooperatives
UCRC	Ugunja Community Resource Centre
UCU	Uganda Christian University
UGX	Uganda Shillings
UNA	National University of Honduras – Universidad Nacional de Agricultura de Honduras
UNBS	Uganda National Bureau of Standards
USD	United States Dollar
VECO	VredesEilanden Country Office
VEDCO	Volunteer Efforts for Development Concern
VIF	variance inflation factor
WeRATE	Western Regional Agricultural Technology Evaluation
WKAS	Western Kenya Action Site
WUR	Wageningen University and Research Center

1 The state of innovation platforms in agricultural research for development

Marc Schut, Jean-Joseph Cadilhon,
Michael Misiko and Iddo Dror

Background

Innovation Platforms (IPs) are widely viewed as a promising vehicle for increasing the impact of agricultural research and development (van Mierlo and Totin, 2014; van Paassen *et al.*, 2014). IPs build on experiences with earlier well-known multi-stakeholder approaches such as Farmer Field Schools (Kenmore *et al.*, 1987; Pontius *et al.*, 2002), Participatory Research (Kerr *et al.*, 2007), Learning Alliances (Lundy *et al.*, 2005; Mvumi *et al.*, 2009), Local Agricultural Research Committees (Hellin *et al.*, 2008) and Natural Resource Management Platforms (Röling, 1994). In the field of agricultural research for development (AR4D), IPs form an important element of a commitment to more structural and long-term engagement between stakeholder groups (Sumberg *et al.*, 2013a). IPs aim to foster agricultural innovation by facilitating and strengthening interaction and collaboration in networks of farmers, extension officers, policy makers, researchers, non-governmental organizations (NGOs), development donors, the private sector and other stakeholder groups. The nature of agricultural innovation can be both technological (e.g. information and communication technology (ICT), agricultural inputs or machinery) and institutional (market approaches, modes of organization, policies and new rules).

An important objective of IPs is to stimulate continuous involvement of stakeholders in describing and explaining complex agricultural problems, and in exploring, implementing and monitoring agricultural innovations to deal with these problems. This is deemed important for three reasons. First, different stakeholder groups can provide various insights about the biophysical, technological and institutional dimensions of the problem, and ascertain what type of innovations are economically, socially, culturally and politically viable (Esparcia, 2014; Schut *et al.*, 2014b). Second, stakeholder groups become aware of their fundamental interdependencies and the need for concerted action to address their constraints and reach their objectives (Leeuwis, 2000; Messely *et al.*, 2013). Third, stakeholder groups are more likely to support and promote specific innovations when they have been part of the decision-making or development process (Faysse, 2006; Neef and Neubert, 2011).

By facilitating interaction between different stakeholder groups, IPs provide space not only for exchange of knowledge and learning (Ngwenya and Hagmann, 2011), but also for negotiation and dealing with power dynamics (Cullen *et al.*, 2014). In so doing, IPs can contribute to strengthening 'capacity to innovate' across stakeholder groups. The capacity to innovate can best be described as the ability of individuals, groups or systems to continuously shape, or adapt to change. This ability stems from varying degrees of resourcefulness in assets, time, knowledge, dialogue, experimentation and persistence. If capacity to innovate is high, individuals, groups and systems are better able to react proactively, flexibly and creatively to shocks, challenges and opportunities (Boogaard *et al.*, 2013a). In summary, an IP's capacity to innovate is related to being able to organize an incentivized process to generate short and long-term benefits for each actor.

In their ability to bring people together, IPs can strengthen capacity to innovate among interdependent groups of stakeholders to:

- continuously identify and prioritize problems and opportunities in a dynamic systems environment;
- take risks, experiment with social and technical options, and assess the trade-offs that arise from these;
- mobilize resources and form effective support coalitions around promising options and visions for the future;
- link with others in order to access, share and process relevant information and knowledge in support of the above;
- collaborate and coordinate with others, and achieve effective concerted action (Leeuwis *et al.*, 2014).

Depending on the specific objective of an IP, and the context in which they function, IPs can operate at different levels. IPs can focus on enhancing the capacity to innovate at the community or village level to address a local productivity problem. However, IPs can also operate at higher levels if the objective is to support the scaling of successful (local) innovations or the facilitation of national policy development and implementation (Cadilhon *et al.*, 2013). If agricultural problems are embedded in interactions and trade-offs across different administrative or spatial levels, interconnected IPs that strengthen the development and implementation of coherent intervention strategies across these different levels may be required (Tucker *et al.*, 2013). Similarly, exploring value-chain innovation through IPs may require the involvement of local producers, regional processors, distributors and retailers, but also of national policy makers and certification bodies (Birachi *et al.*, 2013).

Recent studies on IPs demonstrate their potential in terms of realizing robust agricultural research, development and policy strategies and impact (e.g. Ayele *et al.*, 2012; Kilelu *et al.*, 2013; Schut *et al.*, 2014a; Swaans *et al.*, 2014). However, experiences also show that IPs' performance and impact depend on many variables. For example, the quality of platform organization and facilitation (Rooyen *et al.*, 2013), communication within the IP (Victor *et al.*, 2013), stakeholder representation (Cullen *et al.*, 2013), and institutional embedding determine, to a large extent, whether IPs can lead to real change and impact (Nederlof *et al.*, 2011; Boogaard *et al.*, 2013b; Cullen *et al.*, 2013). Despite all the rhetoric around IPs, there may be an institutional context causing the continuation of 'business as usual' practices, where science develops and tests technologies that are then transferred to end users, often farmers (Friederichsen *et al.*, 2013; Sumberg *et al.*, 2013b; Cullen *et al.*, 2014). Furthermore, several authors have found that resources needed to implement IP approaches are often difficult to obtain in systems that adhere to more traditional linear, top-down approaches to innovation (Kristjanson *et al.*, 2009; Nettle *et al.*, 2013). IPs are not a panacea – a solution to all agricultural problems. There are no blueprints, recipes or silver bullets (Boogaard *et al.*, 2013b), and this is precisely why understanding factors and processes that can contribute to IPs' impact is difficult, but essential.

Documentation of and learning from the effectiveness and impact of IPs is crucial (Lundy *et al.*, 2013). There are many good case studies of IPs published over the past decade (e.g. Nederlof *et al.*, 2011; Nederlof and Pyburn, 2012). However, most, if not all, of these tend to focus on emerging platforms, with limited scale, and a narrow focus (e.g. on a single commodity). With a new 'wave' of IPs in international AR4D, there is a need to reflect on the implementation, sustainability and impact of mature, more established IPs. With this book, we aim to enhance the existing body of knowledge around IPs by focusing on the impact of these mature and established IPs in the AR4D landscape. We realize that many impacts of IPs, such as their contribution to capacity to innovate, are intangible and hard to measure (Boogaard *et al.*, 2013a). There can be time lags between a platform's activities and its impact and it may be difficult to specify the exact contribution of an IP to change or impact (Duncan *et al.*, 2013). Nevertheless it is important to gather evidence about platform actions and achievements, and to speak about and promote successful mature IP case studies.

Case study competition process

Many AR4D programmes, including the CGIAR Research Programs on Integrated Systems for the Humid Tropics (Humidtropics), Climate Change, Agriculture and Food Security (CCAFS), Agricultural Aquatic Systems (AAS), Livestock and Fish, and Maize, as well as the Forum for Agricultural Research in Africa (FARA) Sub-Saharan Africa Challenge Program (SSA CP) have adopted multi-stakeholder approaches to achieve development impacts. Humidtropics, for example, uses integrated systems research and multi-stakeholder approaches to enhance agricultural productivity, eco-systems integrity and institutional innovation. IPs are supposed to drive the demand for concrete research for development activities at the field level, as well as facilitate the active participation of key scaling actors such as the private sector and policy makers at higher levels, where some of the more structural opportunities and constraints for agricultural innovation can be identified.

In 2013, the International Livestock Research Institute (ILRI), as part of its work for Humidtropics, published 12 IP Practice Briefs, intended to inform agricultural research practitioners who seek to support and implement IPs. In the same year, Wageningen University and Research Center (WUR) and ILRI published a Humidtropics paper reviewing critical issues for reflection when designing and implementing Research for Development in IPs (Boogaard *et al.*, 2013b). Several partners also published an IP Guide in 2013, produced through the Kenya Agricultural and Livestock Research Institute (Makini *et al.*, 2013). In April and November 2014, ILRI, WUR and the International Institute for Tropical Agriculture (IITA) organized two Humidtropics workshops in Nairobi, Kenya and Xishuangbanna, China on 'Understanding, Facilitating and Monitoring Agricultural Innovation Processes'. The IP Case Study Competition was launched to continue this quest to decipher the DNA of IPs, and to bring together different stakeholders and actors in the agriculture sector to produce case studies featuring the most innovative ideas, best practices, actionable knowledge and strategies emerging from mature IPs in AR4D.

Contributions to the IP competition were 'crowd-sourced' through an open call for case studies. The theme for the competition was 'Mature innovation platforms in the agricultural systems research landscape'. Under this overarching theme, case studies focused on one of the following topics:

1   *Systems trade-offs*: How have IPs facilitated systems synergies and trade-offs to help farmers maximize production and yield? Trade-offs are a necessary aspect of systems research and agriculture decision making. Analysing system trade-offs helps farmers prioritize their interventions while battling food security, climate change, limited resources, population pressures and technological challenges.

2   *Platforms focusing on multiple commodities*: How have IPs optimized simultaneous work on multiple commodities (e.g. crop–livestock–tree interactions)? Growing more than one kind of crop in the same area – multiple cropping – can help boost the nutrient levels in the soil, protect against harmful weeds, increase the yield of crops and increase revenues from agriculture.

3   *Scaling up agricultural innovations*: How do IPs help scale up agricultural innovations? How have IPs promoted agricultural innovation, the use of new technologies, access to knowledge and markets beyond the initial scope of the platform?

4   *Learning from failure*: 'It's fine to celebrate success but it is more important to heed the lessons of failure' (Bill Gates, www.brainyquote.com/quotes/quotes/b/billgates385735.html). The wisdom of learning from failure is

incontrovertible, yet there are still too few documented cases of the challenges and dynamics that can lead to the failure of platforms.

Applicants were asked to focus on case studies that have a proven impact on a large scale, and that feature mature IPs. Generally, such IPs would have moved beyond the pilot stage and would have had proven results that would be scalable or replicable. Likewise, we encouraged cases that focus on principles, methodologies and ideas that can benefit people everywhere, for example, by highlighting the implementation and role of specific IP concepts (e.g. facilitation, stakeholder representation) in achieving the outcome. During the initial call for case studies, we received 28 abstracts; 7 per cent of the abstracts were submitted under the category systems trade-offs, 32 per cent under the category of multiple commodities, and 46 per cent of the abstracts were submitted under the category scaling up agricultural innovations. None of the abstracts focused on learning from failure. The remaining 15 per cent of the cases were not characterized under one of the specific themes by the authors.
The 28 cases submitted were evaluated for:

- *content strength*: case studies should clearly define the problems and challenges being addressed, construct a detailed and descriptive narrative of how various stakeholders used the IP to create solutions and encourage further thinking and debate on the topic;
- *quality of writing*: case studies should be logically written, with a strong emphasis on good writing and presentation;
- *usefulness of the case study*: case studies should feature only those interventions/programmes that meet the above assessment criteria and have demonstrated long-standing impact. Case studies must feature solutions that are replicable, scalable, sustainable, reliable and relevant for the broader agricultural community.

Based on these evaluation criteria and the four topics, 12 cases were shortlisted after independent review and scoring by the editorial team. The lead authors of these 12 cases were invited to attend a writeshop in Nairobi in February 2015. As part of the preparation process, authors received writing guidelines to draft their case studies. Furthermore, case authors had access to individual mentoring from one of the editors who specialized in case study preparation and creative writing. During the writeshop, participants received training on developing a case outline, telling stories and identifying unique selling points of the case. Furthermore, they could benefit from working with both subject matter experts and communication experts from different CGIAR Centres. Illustrators supported the authors in visualizing their learning experiences.
Following the writeshop, authors had three weeks to finalize and submit their case study. The 12 cases were again reviewed and scored by the editorial team

Table 1.1 Overview of eight case studies that were selected for inclusion in this book (in subsequent chapters we will mainly refer to the short name of the case)

Chapter	Case study title	IP long name	IP short name	Country	Category of submission
2	With trust and a little help from our friends: how the Nicaragua Learning Alliance scaled up training in agribusiness	Nicaragua Learning Alliance	NLA	Nicaragua	3: Scaling up agricultural innovations
3	Overcoming challenges for crops, people and policies in Central Africa – the story of CIALCA stakeholder engagement	Consortium for Improving Agriculture-based Livelihoods	CIALCA	Burundi, DRC and Rwanda	1: Systems trade-offs
4	Can an IP succeed as a cooperative society? The story of Bubaare IP Multipurpose Cooperative Society Ltd	Bubaare Innovation Platform	Bubaare	Uganda	3: Scaling up agricultural innovations
5	Crop–livestock–tree integration in Uganda: the case of Mukono–Wakiso innovation platform	Mukono–Wakiso IP	Mukono–Wakiso	Uganda	2: Platforms focusing on multiple commodities
6	Humidtropics IP case study: WeRATE operations in West Kenya	WeRATE	WeRATE	Kenya	3: Scaling up agricultural innovations
7	IPs for improved natural resource management and sustainable intensification in the Ethiopian Highlands	Humidtropics Ethiopia Local Innovation Platforms	NBDC	Ethiopia	2: Platforms focusing on multiple commodities
8	Sustaining the supply of organic White Gold – the case of SysCom IPs in India	SysCom India	SysCom	India	1: Systems trade-offs
9	MilkIT IP. Changing women's lives – one cow and one litre of milk at a time – deep in the foothills of India's Himalayan mountains	Dairy Value Chain and Feed Innovation Platform	MilkIT	India	3: Scaling up agricultural innovations

independently. Based on the scoring, eight of the 12 cases were found to be suitable for publication in this book (Table 1.1).

Case study characterization and readers' guide

During the writeshop, the editorial team facilitated the participants in several case study characterization exercises that provided more detailed information about the cases. Characterization included their geographical spread, age and life stage, and specific information on the multi-stakeholder processes, the content matter, platform support functions, and outcomes and impacts. Based on the characterization of the case studies, the next section informs readers about the extent to which the different cases address various components of the multi-stakeholder processes: content matter, platform support functions, and outcomes and impacts.

Geographical spread of the case studies

The case studies selected for publication in this compilation cover three continents. One is located in Nicaragua in Central America, while two report experiences from India in Asia. Four cases cover Eastern Africa: one from Ethiopia, one from Kenya and two from Uganda. Finally, one case describes a regional platform covering the three Central African countries of Burundi, the

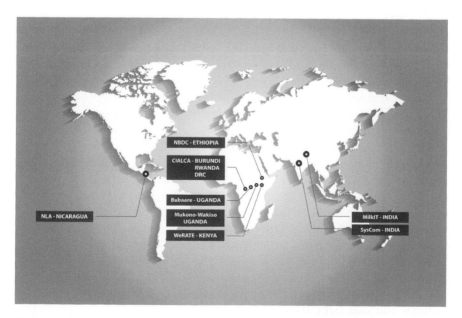

Figure 1.1 World map indicating geographical spread of the eight case studies

Democratic Republic of Congo and Rwanda (Figure 1.1). Cases are ordered by geographical location, from West to East.

Age and life stages of the platforms

The eight IPs featuring in this book vary in the duration of their activities (see Figure 1.2). The youngest platform is the Mukono–Wakiso Humidtropics IP that was only established a year ago. The oldest platform is WeRATE from West Kenya. However, the editors did not consider age as the only criterion of selection for 'mature' platforms. Rather, maturity was approached from a multiple-dimension optic, looking at whether the platforms were embedding multiple commodities, were addressing system trade-offs, or had good inroads in terms of policy impact and scaling. As such, it is more interesting to position the IPs featured along a continuum of 'level of maturity' rather than by the duration of their activities.

IPs generally go through several steps of 'life stages' (Tucker *et al.*, 2013). Their establishment can correspond to their 'birth'. When in 'childhood', IPs concentrate on identifying the problem their members will try to solve collaboratively. The first trials and errors in implementing innovative activities can be linked to an IP's 'adolescence'. The IP can be considered to be in 'adulthood' when its first impacts have been achieved and it starts scaling up its activities for further outreach. When IPs start tackling other R&D problems and strive to scale their innovations further, they have reached 'maturity'. Their mature status can be very long if the IP is considered to be the appropriate tool to keep solving complex multi-stakeholder problems. However, some IPs are also disbanded when they have solved the issue they were meant to address. It also often happens that IPs stop working when external funding dries up and the costs of the meetings and R&D activities cannot be financed internally. This final stage represents the 'death' of the IP.

Despite some of the IPs featuring in this book being relatively young in age, all of them have reached or passed the 'adolescent' stage of trying out innovative activities. Some of the authors of the case studies self-reported their IPs to be at a comparable stage of maturity, even though the platforms had been operating for very different durations. Consider that the Nicaragua Learning Alliance was considered 'adult' by its authors after seven years of activity, whereas the MilkIT platform in India had reached the same life stage after only two years of activity, according to its main author. Likewise, the lead author of the NBDC case from Ethiopia considered that the platform had reached maturity after four years of existence, when it had taken more than ten years of work for Syscom in India and WeRATE in Kenya to reach a similar stage, according to their authors. Finally, the CIALCA IP was also considered to be mature, as the CIALCA stakeholder networks provided the basis for the current Humidtropics work in Burundi, Rwanda and DRC.

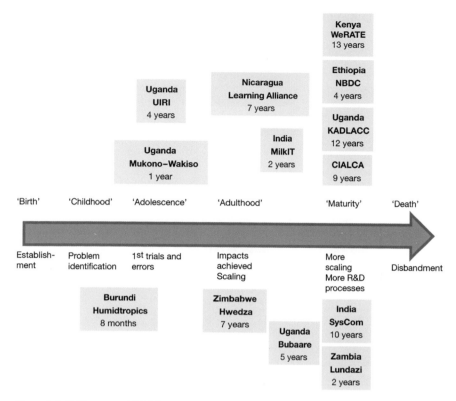

Figure 1.2 Self-reported IP life stages

Platform characteristics

Detailed characterization of the IPs during the writeshop (by the case study authors) and during the assessment of the eight case studies selected during independent review and scoring (by the editorial team) provided a rich picture of the case studies. Focus of the characterizations was put on four interlinked components, namely (1) the multi-stakeholder processes, (2) the content matter, (3) platform support functions and (4) outcomes and impacts.

Figure 1.3 visualizes how these four components are related. It shows how platform support (e.g. facilitation) is required to connect the multi-stakeholder processes of learning, negotiation and experimentation ('how' a problem is identified and addressed) to concrete content matter ('what' is the problem that is bringing together different stakeholders). Outcomes and impacts can both result from the process, as well as from the content matter. An example of process impact could be the strengthening of stakeholder networks, collaboration, interaction and willingness to engage in joint actions. An example of content matter impact could be an innovative seed, breed or any other technology, policy or management practice that is scaled beyond the original scope of the IP.

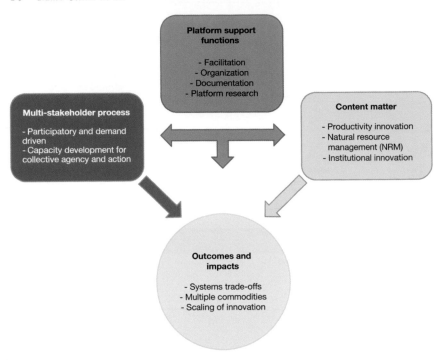

Figure 1.3 Relation between four key components of IP used to characterize the case studies

Multi-stakeholder process

IPs operating in an AR4D context can form an important vehicle for *participatory and demand-driven* research and development activities. Research and development are often disconnected because of the different objectives, time-lines and institutional dynamics of research and development processes. Continuous representation of different groups of stakeholders (including attention for different gender, age and ethnic groups) in research for development (R4D) processes, for example through IPs, can provide better insight into the information, technology and service needs for different groups and their communication and collaboration preferences towards achieving development impact. Furthermore, stakeholders (including politicians, donors and other change agents) are more likely to support and promote specific innovations when they have been part of the innovation and decision-making process (Faysse, 2006). More inclusive and participatory research strategies can support the continuous alignment of research and development strategies with the changing context and stakeholder demands (Greenwood and Levin, 2007). This requires a degree of flexibility and adaptive capacity. The CIALCA, MilkIT and Mukono–Wakiso cases provide good examples of how stakeholder participation and demand-driven R4D can strengthen the contribution of IPs to achieving development impact.

A second key characteristic of multi-stakeholder collaboration in IPs is that they can foster *capacity development for collective agency and action*. Through collaborating in an IP, stakeholder groups can become more aware of their fundamental interdependencies and the need for concerted action to reach their objectives (Leeuwis, 2000). This can provide a basis for better collaboration, investment, joint resource mobilization and policy advocacy. Approaches such as Participatory Learning and Action Research (Wopereis *et al.*, 2007) and Participatory Action Research (e.g. Ottosson, 2003) can provide a good basis for developing the capacity of all involved in IPs. Readers with a specific interest in how IPs can contribute to developing the capacity for collective agency, action and impact are recommended to read the CIALCA, NLA and Bubaare case studies.

Content matter

To assess and categorize the content matter addressed in the IP case studies, we look at three types of agricultural innovations. The first one deals with novel technologies and management practices to *increase productivity* (based on laboratory and field science). Readers with an interest in productivity innovation should definitely have a look at the CIALCA, SysCom, WeRATE and Mukono–Wakiso cases. The second type of innovations are related to responsible NRM that deal with low soil fertility, low yields, erosion, deforestation and climate change (Misiko *et al.*, 2013). The NBDC and SysCom cases deal with such NRM innovations. The third type of innovations are geared towards creating an enabling institutional environment (or *institutional innovation*) that can include: enhanced collaboration between stakeholders, social infrastructure, access to finance, certification, land tenure arrangements, and public goods and markets (Pretty *et al.*, 2011). Of the case studies included in this book, the CIALCA, MilkIT and Bubaare cases provide good examples of how IPs can contribute to institutional innovation.

An important element of systems approaches is that productivity, NRM and institutional innovations need to emerge in an integrated way, making smart use of available agro-ecological and human resources across different systems levels (Robinson *et al.*, 2015). Both the CIALCA and SysCom cases address two of the three types of innovation.

Platform support functions

Effective support to, and learning from, multi-stakeholder processes in agricultural R4D interventions requires four major critical success functions. The first one, *facilitation*, is usually fulfilled with a small team of people. Facilitation refers to ensuring sufficient linkages and empowerment of the process participants. The linkages not only cover the connections between participants but also those of the IP with markets, donors and political decision makers (Rooyen *et al.*, 2013). Facilitation, and how it contributed to platform impact, is described in depth in the Mukono–Wakiso case from Uganda. The second critical success function is *organization*. Organization refers to provision of logistical support, backstopping of events and administering the accountability work. Typical examples are renting the venue, providing lunch and handling IP finances. The Ugandan Bubaare and Mukono–Wakiso cases stand out in terms of their reflection on platform organization. The third function is *documentation*. Documentation refers to the systematic capturing and reporting of events and developments in the process. Documentation and learning systems should be inclusive and participatory. IP members should participate in monitoring, and information should be gathered continuously and fed back quickly. As such, the monitoring and learning system becomes a tool for reflection on both the platform process and its ability to develop solutions to concrete problems (Lundy *et al.*, 2013). Readers with a particular interest in documentation of platform process and impact should definitely have a look at the NLA and Mukono–Wakiso cases. Lastly, *research* on the platform process function is critical. In the existing international AR4D landscape, sufficient prioritization of the learning tasks and funding of the learning activities are highly correlated with the availability and enthusiasm of the researchers championing process research (Lema and Schut, 2013). Platform research receives particular attention in the Mukono–Wakiso case, which stands out overall in terms of its attention to platform support functions.

Outcomes and impact

When categorizing the case studies, editors assessed the outcomes and impacts of the platforms under four categories. The first category is *systems trade-offs*, exploring synergies and competition between different interventions and strategies. Trade-offs can be of financial (where to invest in?), social (how to allocate labour?) or technological (mono- versus inter-cropping?) nature. The NBDC case provides some very good examples of how IPs can support optimization of systems trade-offs. The second category of impacts is IPs focusing on *multiple commodities*, for example on managing complex crop–livestock–tree interactions. WeRATE from Kenya, and Bubaare and Mukono–Wakiso from Uganda provide good examples. The third category of outcomes and impacts are related to the *scaling up of agricultural innovations*. Scaling relates to the use of new technologies, dissemination of (scientific) knowledge, collaborations between different stakeholder groups, access to markets, etc. beyond the original IP scope, geographical focus or target audience. Readers interested in learning more about how platforms can reach impact at scale should have a close look at the WeRATE case. As explained, no cases were submitted under the fourth category of *learning from failure*.

Book outline

The eight following chapters are the case studies of mature IPs selected by the editors from contributions to the competition. Readers are invited to refer to the readers' guide above to identify which case studies are more likely to tackle their area of interest along the four components of multi-stakeholder process, content matter, platform support functions, and outcomes and impact.

Chapter 10 provides a synthesis of the key relations and impact pathways that exist between the three components of IPs and outcomes and impacts, as illustrated by the eight case studies featured in this book. The conclusion of the book provides lessons learned from the case studies on how to implement IPs that will deliver impact. It also reflects on the current landscape of mature IPs and tries to answer the question of whether IPs have managed to achieve impact at scale in agricultural development.

References

Ayele, S., Duncan, A., Larbi, A., Khanh, T.T., 2012. Enhancing innovation in livestock value chains through networks: Lessons from fodder innovation case studies in developing countries. *Science and Public Policy* 39, 333–346.

Birachi, E., Rooyen, A.v., Some, H., Maute, F., Cadilhon, J., Adekunle, A., Swaans, K., 2013. Innovation platforms for agricultural value chain development. Innovation Platforms Practice Brief 6. ILRI, Nairobi, Kenya.

Boogaard, B., Dror, I., Adekunle, A., Borgne, E.L., Rooyen, A.v., Lundy, M., 2013a. Developing innovation capacity through innovation platforms. Innovation Platforms Practice Brief 8. ILRI, Nairobi, Kenya.

Boogaard, B., Klerkx, L., Schut, M., Leeuwis, C., Duncan, A., Cullen, B., 2013b. Critical issues for reflection when designing and implementing Research for Development in innovation platforms. Report for the CGIAR Research Program on Integrated Systems for the Humidtropics. Knowledge, Technology & Innovation Group (KTI), Wageningen University and Research Centre, Wageningen, the Netherlands, p. 42.

Cadilhon, J., Birachi, E., Klerkx, L., Schut, M., 2013. Innovation platforms to shape national policy. Innovation Platforms Practice Brief 2. ILRI, Nairobi, Kenya.

Cullen, B., Tucker, J., Tui, S.H.-K., 2013. Power dynamics and representation in innovation platforms. Innovation Platforms Practice Brief 4. ILRI, Nairobi, Kenya.

Cullen, B., Tucker, J., Snyder, K., Lema, Z., Duncan, A., 2014. An analysis of power dynamics within innovation platforms for natural resource management. *Innovation and Development* 4, 259–275.

Duncan, A., Borgne, E.L., Maute, F., Tucker, J., 2013. Impact of innovation platforms. Innovation Platforms Practice Brief 12. ILRI, Nairobi, Kenya.

Esparcia, J., 2014. Innovation and networks in rural areas. An analysis from European innovative projects. *Journal of Rural Studies* 34, 1–14.

Faysse, N., 2006. Troubles on the way: An analysis of the challenges faced by multi-stakeholder platforms. *Natural Resources Forum* 30, 219–229.

Friederichsen, R., Minh, T.T., Neef, A., Hoffmann, V., 2013. Adapting the innovation systems approach to agricultural development in Vietnam: Challenges to the public extension service. *Agriculture and Human Values* 30, 555–568.

Greenwood, D.J., Levin, M., 2007. *Introduction to Action Research. Social Research for Social Change*, 2nd edn. Sage, Thousand Oaks, CA.

Hellin, J., Bellon, M.R., Badstue, L., Dixon, J., La Rovere, R., 2008. Increasing the impacts of participatory research. *Experimental Agriculture* 44, 81–95.

Kenmore, P., Litsinger, J.A., Bandong, J.P., Santiag, A.C., Salac, M.M., 1987. Philippine rice farmers and insecticides: Thirty years of growing dependency and new options for change, in: Tait, J., Napompeth, P. (eds), *Management of pests and pesticides: Farmers' perceptions and practices*. Westview Press, Boulder, CO, pp. 98–115.

Kerr, R.B., Snapp, S., Chirwa, M., Shumba, L., Msachi, R., 2007. Participatory research on legume diversification with Malawian smallholder farmers for improved human nutrition and soil fertility. *Experimental Agriculture* 43, 437–453.

Kilelu, C.W., Klerkx, L., Leeuwis, C., 2013. Unravelling the role of innovation platforms in supporting co-evolution of innovation: Contributions and tensions in a smallholder dairy development programme. *Agricultural Systems* 118, 65–77.

Kristjanson, P., Reid, R.S., Dickson, N., Clark, W.C., Romney, D., Puskur, R., MacMillan, S., Grace, D., 2009. Linking international agricultural research knowledge with action for sustainable development. *Proceedings of the National Academy of Sciences* 9, 5047–5052.

Leeuwis, C., 2000. Reconceptualizing participation for sustainable rural development: Towards a negotiation approach. *Development and Change* 31, 931–959.

Leeuwis, C., Schut, M., Waters-Bayer, A., Mur, R., Atta-Krah, K., Douthwaite, B., 2014. Capacity to innovate from a system CGIAR research program perspective. Penang, Malaysia: CGIAR Research Program on Aquatic Agricultural Systems. Program Brief: AAS-2014-29, p. 5.

Lema, Z., Schut, M., 2013. Innovation and research platforms. Innovation Platforms Practice Brief 3. ILRI, Nairobi, Kenya.

Lundy, M., Gottret, M.V., Ashby, J., 2005. Learning alliances: An approach for building multi-stakeholder innovation systems, ILAC Brief 8. Bioversity, Rome.

Lundy, M., LeBorgne, E., Birachi, E., Cullen, B., Boogaard, B., Adekunle, A., Victor, M., 2013. Monitoring innovation platforms. Innovation Platforms Practice Brief 5. ILRI, Nairobi, Kenya.

Makini, F.W., Kamau, G.M., Makelo, M.N., Adekunle, W., Mburathi, G.K., Misiko, M., Pali, P., Dixon, J., 2013. *Operational field guide for developing and managing local agricultural innovation platforms*. KARI and ACIAR, Nairobi.

Messely, L., Rogge, E., Dessein, J., 2013. Using the rural web in dialogue with regional stakeholders. *Journal of Rural Studies* 32, 400–410.

Misiko, M., Mundy, P., Ericksen, P., 2013. Innovation platforms to support natural resource management. Innovation Platforms Practice Brief 11. ILRI, Nairobi, Kenya.

Mvumi, B.M., Morris, M., Stathers, T.E., Riwa, W., 2009. Doing things differently: Post-harvest innovation learning alliances in Tanzania and Zimbabwe, in: Sanginga, P.C., Waters-Bayer, A., Kaaria, S., Njuki, J., Wettasinha, C. (eds), *Innovation Africa, enriching farmers' livelihoods*. Earthscan, London, p. 199.

Nederlof, E.S., Pyburn, R., 2012. *One finger cannot lift a rock: Facilitating innovation platforms to trigger institutional change in West Africa*. Royal Tropical Institute, Amsterdam.

Nederlof, S., Wongtschowski, M., van der Lee, F., 2011. Putting heads together: Agricultural innovation platforms in practice, *Bulletin 396*. KIT Publishers, Amsterdam, the Netherlands, p. 192.

Neef, A., Neubert, D., 2011. Stakeholder participation in agricultural research projects: A conceptual framework for reflection and decision-making. *Agriculture and Human Values* 28, 179–194.

Nettle, R., Brightling, P., Hope, A., 2013. How programme teams progress agricultural innovation in the Australian dairy industry. *Journal of Agricultural Education and Extension* 19, 271–290.

Ngwenya, H., Hagmann, J., 2011. Making innovation systems work in practice: Experiences in integrating innovation, social learning and knowledge in innovation platforms. *Knowledge Management for Development Journal* 7, 109–124.

Ottosson, S., 2003. Participation action research: A key to improved knowledge of management. *Technovation* 23, 87–94.

Pontius, J., Dilts, R., Bartlett, A., 2002. Ten years of building community: From Farmer Field Schools to community IPM. Community IPM Programme, FAO, Jakarta, Indonesia.

Pretty, J., Toulmin, C., Williams, S., 2011. Sustainable intensification in African agriculture. *International Journal of Agricultural Sustainability* 9, 5–24.

Robinson, L.W., Ericksen, P.J., Chesterman, S., Worden, J.S., 2015. Sustainable intensification in drylands: What resilience and vulnerability can tell us. *Agricultural Systems* 135, 133–140.

Röling, N., 1994. Platforms for decision making about ecosystems, in: Fresco, L.O., Stroosnijder, L., Bouma, J., van Keulen, H. (eds), *The future of the land: Mobilizing and integrating knowledge for land use options*. John Wiley & Sons, Chichester, pp. 385–393.

Rooyen, A.v., Swaans, K., Cullen, B., Lema, Z., Mundy, P., 2013. Facilitating innovation platforms. Innovation Platforms Practice Brief 10. ILRI, Nairobi, Kenya.

Schut, M., Cunha Soares, N., van de Ven, G.W.J., Slingerland, M., 2014a. Multi-actor governance of sustainable biofuels in developing countries: The case of Mozambique. *Energy Policy* 65, 631–643.

Schut, M., van Paassen, A., Leeuwis, C., Klerkx, L., 2014b. Towards dynamic research configurations: A framework for reflection on the contribution of research to policy and innovation processes. *Science and Public Policy* 41, 207–218.

Sumberg, J., Heirman, J., Raboanarielina, C., Kaboré, A., 2013a. From agricultural research to 'product development': What role for user feedback and feedback loops? *Outlook on Agriculture* 42, 233–242.

Sumberg, J., Thompson, J., Woodhouse, P., 2013b. Why agronomy in the developing world has become contentious. *Agriculture and Human Values* 30, 71–83.

Swaans, K., Boogaard, B., Bendapudi, R., Taye, H., Hendrickx, S., Klerkx, L., 2014. Operationalizing inclusive innovation: Lessons from innovation platforms in livestock value chains in India and Mozambique. *Innovation and Development* 4, 239–257.

Tucker, J., Schut, M., Klerkx, L., 2013. Linking action at different levels through innovation platforms? Innovation Platforms Practice Brief 9. ILRI, Nairobi, Kenya.

van Mierlo, B., Totin, E., 2014. Between script and improvisation: Institutional conditions and their local operation. *Outlook on Agriculture* 43, 157–163.

van Paassen, A., Klerkx, L., Adu-Acheampong, R., Adjei-Nsiah, S., Zannoue, E., 2014. Agricultural innovation platforms in West Africa: How does strategic institutional entrepreneurship unfold in different value chain contexts? *Outlook on Agriculture* 43, 193–200.

Victor, M., Ballantyne, P., Borgne, E.L., Lema, Z., 2013. Communication in innovation platforms. Innovation Platforms Practice Brief 7. ILRI, Nairobi, Kenya.

Wopereis, M.C.S., Defoer, T., Idinoba, M.E., Diack, S., Dugué, M.J., 2007. *Participatory learning and action research (PLAR) for integrated rice management (IRM) in inland valleys of sub-Saharan Africa: Technical manual*. WARDA (Africa Rice Centre), Cotonou, Benin/ IFDC, Muscle Shoals, USA.

2 With trust and a little help from our friends

How the Nicaragua Learning Alliance scaled up training in agribusiness

*Dirk Hauke Landmann and
Jean-Joseph Cadilhon*

Will Nicaragua become the next basket case of failed agricultural development? Unfortunately, trends do not look promising. Nicaragua is the second poorest and one of the least developed countries in Latin America (The World Bank Group, 2014b). Its development story has gone through natural hazards and major upheavals in its society and political system. As a result, 42 per cent of the 6.08 million Nicaraguan population is still living in rural areas in 2013 (FAOSTAT, 2014) and 80 per cent of the poor live in the countryside (The World Bank Group, 2014a). Although agriculture is a main driver of economic growth, representing 22 per cent of Nicaraguan GDP, it is characterized by low productivity (FAOSTAT, 2014). The government has tried to strengthen the economy over the past 20 years by increasing exports and foreign direct investments but the strategy was not successful due to the 2008–2009 global financial crisis (The World Bank Group, 2014a). Furthermore, Nicaraguan farmers are generally not aware of business entrepreneurship and market dynamics (CATIE, 2008). Not being able to link themselves to markets or to build a robust business plan put farmers in weak positions when doing business with their input suppliers and produce buyers. It is this last challenge that partners involved in the Nicaragua Learning Alliance are trying to address.

On the brighter side, the Nicaraguan agricultural sector is well organized: 4,124 cooperatives were operating on agricultural topics in 2007, representing 62 per cent of all cooperatives in the country. They were spread out to cover all agricultural products and provinces (Lafortezza and Consorzio, 2009). The Nicaragua Learning Alliance (NLA) is a national IP that was founded in 2008. It has been able to leverage this dense network of cooperatives to strengthen the awareness of farmers' organizations and their members on agribusiness development in all types of agricultural products. Overall, the ten NLA members have trained representatives in 77 producers' cooperatives, who then trained a total of 19,347 households in Nicaragua thanks to a snowball training mechanism, the trust developed in the project managers and the relevance of their training methods. Our case findings also show that the cooperatives trained by the NLA do recognize the Alliance, rather than other agribusiness training networks, as the provider of the applicable knowledge and skills they have learned. This case study uncovers how the NLA has organized its training process to reach so many final beneficiaries, and evaluates the alliance's setup in view of its expected outcomes in knowledge development.

More efforts needed to develop Nicaragua's agribusiness base

Agriculture accounts for 32 per cent of Nicaragua's exports and 32 per cent of its employment (Lafortezza and Consorzio, 2009). The agricultural labour force is dominated by men (92 per cent). Coffee is the most economically important product in the country's otherwise diversified agricultural production (Table 2.1). Coffee is also the product with the biggest export value, followed by beef, sugar, peanuts and milk products (FAOSTAT, 2014).

The agricultural sector has been heavily influenced by the country's turbulent history. The year 1979 marked the triumph of the Sandinista revolution, and the beginning of socialist reforms in which land distribution played a central role. Soon after taking power, the Sandinista government began seizing large farms and redistributing land among rural landless poor and organizing farmers into cooperatives.

However, the Revolution was short-lived and the socialist regime was replaced by a market-oriented government after just ten years. Consequently, many agricultural cooperatives were dissolved and farmers began cultivating their land individually. Nevertheless, many cooperatives still exist (Ruben and Lerman, 2005). Cooperatives are also geographically widely spread across the

Figure 2.1 Jesús Matamoros, smallholder coffee producer on 'El Plan' farm, community of Las Escaleras, Matagalpa, Nicaragua

Photo: CIAT/Adriana Varón

Table 2.1 Nicaraguan principal agricultural products and their share of agricultural GDP

Product	Percentage of total agricultural GDP
Coffee	20
Beans	14
Sugar cane	11
Maize	9
Rice	9
Nuts	7
Others	30

Source: Lafortezza and Consorzio, 2009

country (Lafortezza and Consorzio, 2009). Farmers have numerous motives for participating in these cooperatives: access to financial support and credit, extension agents, etc. According to the Central American Bank for Economic Integration (BCIE) (Table 2.2), there were 6,655 cooperatives in Nicaragua in 2007, 62 per cent of which were in the agricultural sector (Lafortezza and Consorzio, 2009).

Figure 2.2 Smallholder coffee producer José Pérez, his wife Gloria with children and grandchildren, 'La Loma' farm, community of Las Escaleras, Matagalpa, Nicaragua

Photo: CIAT/Adriana Varón

Table 2.2 Registered cooperatives in Nicaragua in 2007

Sector	Total	(%)
Agriculture	4,124	61.97
Transport	966	14.52
Multiple services	454	6.82
Fishery	366	5.50
Savings and credits	323	4.85
Multisectorial	106	1.59
Others	316	4.75
Total	6,655	100

Source: Lafortezza and Consorzio, 2009

Nicaraguan agriculture still has a significant potential to increase its production. This is particularly important considering agriculture is a major driver of the economy, both domestically and through exports. The government is targeting smallholders like José Pérez and his family (see Figure 2.2) because they produce most of the country's agricultural goods (The World Bank Group, 2012). Smallholder farmers in Nicaragua are still facing technical hurdles such as access to water and battling crop and livestock diseases, which lead to low productivity (CATIE, 2008). This low productivity in turn hinders public and private investments, technological innovation, business development services and access to rural finance. The socialist past also explains how Nicaraguan farmers and their organizations have rather weak skills in agribusiness management and development. As a result, they are not well equipped to link themselves to suppliers and customers in today's market-oriented system. International development partners such as CIAT, CARE, CRS and others realized that agribusiness training would be a better long-term strategy to empowering rural farming communities in Latin America than showering aid money on them. They thus created the regional Learning Alliance (LA)[1] for Latin America to foster agribusiness training among Latin American smallholder farmers. The Nicaraguan partners of the LA then went on to set up the NLA to reach this regional objective in Nicaragua (Lundy and Gottret, 2005).

How the NLA trained over 19,000 farmer households from beach to mountain in Nicaragua

Organization of the Learning Alliance

The development partners who were members of the regional learning alliance met to identify the topics for learning that would be relevant for most countries where they had activities in Latin America. Having identified agribusiness development as a useful training topic to empower smallholder farmers and their organizations, they developed a standardized training method that was then used in the different national platforms. The methodology utilizes an approach for

Figure 2.3 Cover pages of NLA guide no. 1 on self-evaluation for the management of rural associative enterprises and guide no. 2 on strengthening socio-organizational processes in farmers' groups

Source: CATIE, www.catie.ac.cr/es/

strengthening the socio-organizational and business management of rural agricultural enterprises. It includes a series of five methodological and training guides covering several topics (AdA, 2014a). The first two guides focus on the organizational skills of farmers' groups: self-evaluation provided for the management of rural associative enterprises and strengthening farmers' groups' socio-organizational processes (see Figure 2.3). The third and fourth training guides aim to deal with managing an agribusiness enterprise: strategic orientation with a focus on value chain and business plans development. Finally, the fifth guide targets farmers' organizations with training on strengthening of services.

The process of each learning alliance is structured in cycles (Figure 2.4) in which the alliance members and their partners follow the process along seven steps (AdA, 2014b):

1 identify what stakeholders want to learn at the end of the process (question of learning);
2 recognize the knowledge that currently exists that could provide an answer to the question (a good existing practice);
3 select the methods or tools identified as good practices to use or adapt (prototype) to answer the question of learning;
4 co-develop the prototype in practice that applies in the field, through training and personal guidance;
5 implement the developed prototype (field application);

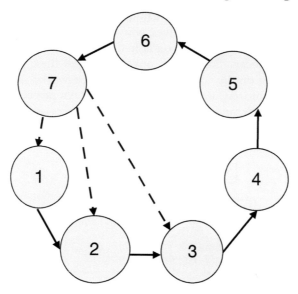

Figure 2.4 Learning cycle of the Learning Alliances
Source: AdA, 2014b

6   write workshops to reflect on the lessons learned and share the results with others (documentation and systematization of results);
7   identify empirical evidence for the conceptual development and recognize political implications, which will lead to improved practices and knowledge (selection of learning).

In Nicaragua a number of different NGOs came together to form the NLA with the International Center for Tropical Agriculture (CIAT). These included CRS, FUNICA, GIZ, LWR, OXFAM, SwissContact and VECO Mesoamerica. They were joined by CATIE, a research organization and FENACOOP R.L., a third-level national farmers' cooperative. The NLA completed three learning cycles between 2008 and 2013, with training activities and beneficiaries concentrated in the provinces of Matagalpa, Jinotega, Estelí, Madriz and Nueva Segovia (Figure 2.5).

The NLA used the dense network of NGOs and farmers' cooperatives in Nicaragua (Figure 2.6) to scale up its training on agribusiness development.

The NLA members listed above first constituted a working group. Each of the NLA members assigned and sent a representative, the project manager, who worked actively in the group to develop and improve training guides. The project managers then used these guides at the provincial offices of their organizations to train second-level cooperatives: unions or associations of farmers' cooperatives that operate at the local level in a given province.

This chain continued further, with second-level cooperatives training representatives of first-level cooperatives, who represent producers in rural areas. Finally, the first-level cooperatives replicated the training for their members: the individual producers. Sometimes, one or more of these levels would be skipped, depending on the configuration of local networks. To improve the guides during the process described above, the NLA's project managers had regular meetings to exchange information and experiences on how the trainings went.

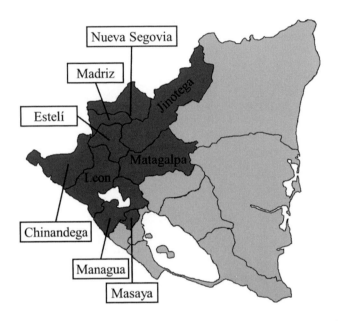

Figure 2.5 Provinces of Nicaragua where data collection occurred for this study

Source: Own graphic

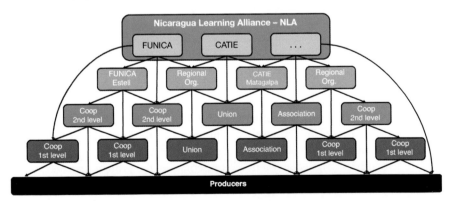

Figure 2.6 Structure of training process within the NLA

Source: Own research

NLA results: more than 19,000 farming households benefitted from agribusiness training

Because training is a development intervention with longer-term impacts than direct aid to help beneficiaries take active decisions on improving their lives, the NLA placed training at the forefront of its strategy and committed massive financial resources for it. The NLA members initially contributed USD341,740 to developing the first two learning cycles between 2008 and 2012. They also directly invested money to support 77 participating farmers' organizations. The first learning cycle included 26 producers' organizations and reached a total of 6,647 farming families producing coffee, cocoa, vegetables, basic grains, plantains, roots and tubers, milk and honey. Some 30 per cent of these participants and partners were women. The second and third learning cycles covered another 51 producers' organizations, representing around 12,700 families producing coffee, cocoa, vegetables, basic grains, dairy, honey, rice, banana, sugarcane, sesame and cashew nuts (AdA Nicaragua, 2012).

Some NLA members are still using the guides to train their partners outside of the official NLA learning cycles. The NLA distributed self-evaluation forms allowing every farmer who used the guides to measure his or her business against the status quo and detect the areas in which opportunities exist for improvements. CATIE also published a book in 2010 with reports from 23 partners participating in the NLA activities (Lorio *et al.*, 2010). It documents the success of the LA method in Nicaragua with respect to the guides used.

The NLA was thus successful in training a large number of individual Nicaraguan farmers by using the dense network of agricultural cooperatives, to which a majority of farmers are affiliated (Figure 2.6). But the question still remains: did all this training by the NLA and its network of participating cooperatives contribute to real agribusiness development of smallholder farmers? If yes, then how did this impact come about?

Research model and method to understand how IPs work

To understand how the NLA works and how it manages, or not, to reach expected training outcomes, this case study combines three different approaches to form one model (Cadilhon, 2013). The overall logic of the model is borrowed from the Structure–Conduct–Performance (SCP) Model coming from industrial organization theory. Applied to IPs, our model assumes that the structure of the platform impacts the conduct of its members that in turn impacts the performance of the platform.[2] In other words, how an IP is organized directs how its members interact and do business together, which over time determines how successful the IP is at fulfilling its objectives. Our model also borrows some insights from New Institutional Economics. This theory recognizes the existence of complex and sometimes nebular types of multi-stakeholder entities (platforms, groups, institutions, organizations) within societies and markets.

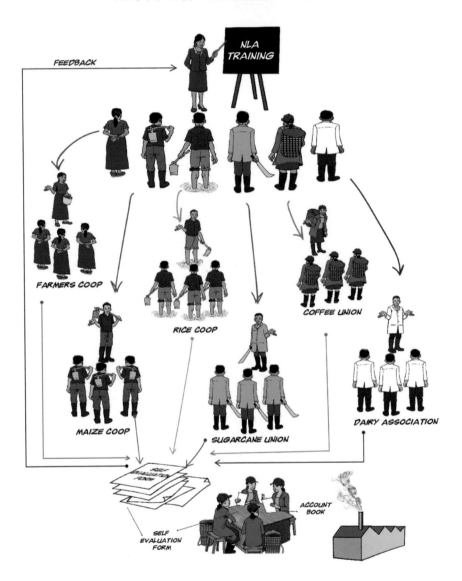

NICARAGUA LEARNING ALLIANCE

NLA TRAINING

FEEDBACK

FARMERS COOP

RICE COOP

COFFEE UNION

MAIZE COOP

SUGARCANE UNION

DAIRY ASSOCIATION

SELF EVALUATION FORM

ACCOUNT BOOK

SELF EVALUATION FORM

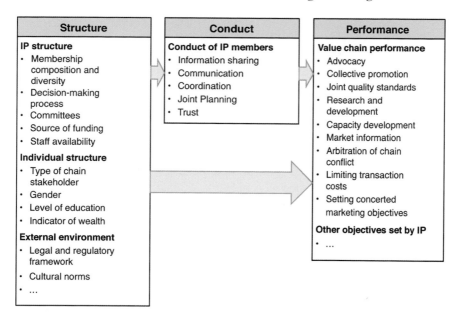

Figure 2.7 Elements of a theoretical model to monitor and evaluate the performance of IPs in a value chain context

Source: Cadilhon, 2013

Finally, the variables used to measure structure, conduct and performance in our model are adapted from the marketing research and business relationships literature to fit IPs (Figure 2.7).

In this model, some elements characterize how IP members act. These elements are defined as information sharing, communication, coordination, joint planning and trust. The elements characterizing the performance of national-level platforms such as the NLA are advocacy, value chain development, nurturing smaller platforms and capacity development (Cadilhon, 2013).

Although three-quarters of the survey respondents were men, the total farmer membership of the organizations the respondents represented was made up of 69 per cent men and 31 per cent women producers. Three cooperatives interviewed were women-only; all the others were mixed-gender cooperatives. Appendix 2.1 shows the main characteristics of the individual survey respondents and the farmers' organizations they represent. It is worth highlighting two points: the majority of farmers' organizations were involved in several agricultural products and the most important source of funding for the respondents' organizations came from NGOs. Appendix 2.2 details all the descriptive tables of the quantitative data we collected.

Box 2.1 Research methodology

In this study, we concentrate on trust as the indicator of platform conduct. Our analysis also focuses on capacity development to evaluate the NLA's training performance. We gathered both qualitative and quantitative data. We interviewed 20 key informants, held five focus group discussions with individual farmers and observed meetings of various actors in the agricultural sector (Landmann, 2015). By mixing introductions from NLA members, random sampling and snowball sampling, we managed to complete 90 individual surveys: 38 respondents represented a farmers' organization involved in the NLA network; another 52 representatives of farmers' organizations not involved in the NLA network represented our control group for the quantitative data. We then analysed the quantitative data using descriptive statistics procedures, analysis of variance, factor analysis and regression analysis. We used the qualitative data to triangulate the results from the statistical analyses so as to validate our theoretical model.

The Central American Bank for Economic Integration (BCIE) undertook a comparable study of Central American farmers' organizations. It collected data from 63 representative Nicaraguan cooperatives (Lafortezza and Consorzio, 2009). Our data sample shows similar results to the BCIE study in terms of main commodities produced and exported by the cooperatives, and in terms of gender balance in the farmers' organizations. Differences are found mainly in the size of the interviewed organizations whereby our study also includes some second-level and first-level cooperatives with more than 10,000 members. Moreover, 35 per cent of the BCIE sample had not received any training whereas all the farmers' organizations in our sample were connected to a training partner. Despite these differences, the overall similarities allow us to consider that our sample is representative of the farmers' organizations in the provinces where the NLA is active.

The NLA is as good as other networks in agribusiness training to farmers

Statistically speaking, there was no significant difference in the conduct and performance of NLA network members as compared with the control group. Thus, despite all the money and efforts invested by the NLA members into the learning alliance, participating in the NLA learning cycles did not give beneficiaries an advantage in strengthening interactions between network partners nor in improving skills in agribusiness management.

The reason for this lack of difference is the current structure of Nicaraguan agriculture. Agricultural cooperatives are a very common way for farmers to organize themselves. Many farmers are members of more than one cooperative undertaking different activities: e.g., financial support or credit, production of different agricultural goods, multi-sectorial cooperatives. All the first-level cooperatives interviewed were working directly or indirectly with other partners such as second-level cooperatives, third-level cooperatives, national associations, unions, farmers' field schools, NGOs, research institutes, private sector players such as traders, exporters or processors, and with governmental institutions such as INTA, MAGFOR or MEFCCA. All key informants and farmers involved in the focus group discussions confirmed they participated in more than one organization conducting training. Furthermore, 78 per cent of individual respondents said their organization was participating in more than one group or learning network.

Because of the abundant supply of agricultural development partners, the NLA was not alone in training farmers nor did it impact their behaviour significantly. So were the NLA's massive funding and intensive activities a waste? The next section reveals the costly though intangible factor that cements the entire learning alliance network together and contributes to its success: trust.

Foster trust for long-term success in agribusiness training

Trust building is a complex and integral part of sustainable business relationships (Laeequddin *et al.*, 2010). Trust is often fostered by many components and actions of the business partners such as regular physical and institutional interactions, expectations fulfilled, a recognized brand name, a written contract. Although the NLA is not a supplier–customer business relationship, it is attempting to develop the agribusiness mentality of beneficiary organizations, so it is relevant to study effects of trust in this IP (Cadilhon, 2013).

The relationship built between NLA members and their network of farmer's organizations often consisted of more than just the training guides of the learning cycles: there was also co-funding and other technical training provided. All these other activities and more frequent physical meetings with the project manager from the NLA members contributed to building up the trust between the NLA members and the farmers' organizations they work with directly. This can help explain the findings reported in Tables 2.3 and 2.4. Although there was no statistically significant difference in the NLA's overall performance on agribusiness training, there were significant differences when going deeper into the local networks involved.

Representatives of farmers' cooperatives active at the second level of the network were getting training directly from the project manager of the NLA member. They tended to agree more that their knowledge and skills in agribusiness had improved thanks to their connection to the NLA when compared with representatives of farmers' cooperatives in a similar position within the network but who were not being trained by an NLA member. First-level cooperatives, who got trained by the second-level cooperatives rather than by

Table 2.3 Appreciation of capacity development performance by second-level cooperatives

Level	Second-level cooperative	
Element	Performance–capacity development	
Statement	6. In the past six years, we have gained knowledge and skills applicable in my activities from NLA stakeholders.★	
NLA-connection	Not a member/no connection	Member/connection
Mean★	2.40	4.43
Standard deviation	1.52	.53

★ Scale: 1 = strongly disagree; 2 = disagree; 3 = undecided; 4 = agree; 5 = strongly agree. Means are statistically significantly different at a 1% level.

Source: Own data collection and analysis

Table 2.4 Appreciation of capacity development performance by first-level cooperatives

Level	First-level cooperative	
Element	Performance–capacity development	
Statement	6. In the past six years, we have gained knowledge and skills applicable in my activities from NLA stakeholders.★	
NLA-connection	Not a member/no connection	Member/connection
Mean★	3.50	4.42
Standard deviation	1.73	.58

★ Scale: 1 = strongly disagree; 2 = disagree; 3 = undecided; 4 = agree; 5 = strongly agree. Means are statistically significantly different at a 5% level.

Source: Own data collection and analysis

the project manager from an NLA member, reported a smaller effect of the NLA on their improved knowledge and skills in agribusiness since the NLA activities started.

Short reckonings make long friends: satisfactory financial dealings also helped sustain trust between partners. Many cooperatives interviewed saw financial support as a basic need that had to come with technical training to be successful, reflecting the fact that the majority of organizations interviewed had NGOs as their main source of funding. This has to be taken critically because financial support should not be indefinite. Rather, the main objective is to have successful producers' groups that are not overly dependent on external financial support (Lundy and Gottret, 2005).

One important element of trust in a business relationship is the personal relationship developed between representatives of organizations doing business together. The effect of having a dedicated project manager involved in the relationship on trust building is illustrated by the counter-example of one of the NLA members: the national-level farmers' cooperative FENACOOP. Like the other NLA members, FENACOOP duly appointed a project manager in charge of representing the cooperative and working with the NLA. However, due to financial issues, the project manager was made redundant and nobody took over his tasks. The cooperative had to leave the NLA in the middle of a learning cycle and it discontinued teaching the modules to its network of local-level farmers' cooperatives. Conversely, the NGO FUNICA and the local research institute CATIE were some of the most active members of the NLA and adopted and extended all the guides within their networks. FUNICA's and CATIE's project managers worked very closely with their clients whether they were in a learning cycle or not. As FENACOOP stopped teaching the guides to their partners the trust in FENACOOP did not increase and the knowledge about agribusiness through the guides did not improve within its network. These are the reasons why the FENACOOP partners disagreed with statements related to 'increased trust in NLA products' (Table 2.5) and 'NLA's success' (Table 2.6) when compared with cooperatives working with other NLA members.

Finally, trust is often built from seeing expectations and commitments delivered by the business partner. Farmers in the focus group discussions said that they had more trust in the NGOs than in the government because the former were more reliable and had more financial resources that could be given to the cooperatives (INTA, 2011).

We also imputed our quantitative data in a regression analysis to show how structure variables had an impact on developing trustful relationships (Table 2.7). The regression results confirmed that the proximity of the farmers' organization with the NLA member within the network has an impact on trust: cooperatives active in the network at national and regional levels have

Table 2.5 Appreciation of trust on products provided by different NLA members

Element	Conduct–trust	
Statement	8. Our trust on products provided by the NLA/our organization has increased.★	
NLA-member	Mean	Standard deviation
FUNICA	4.21	.70
CATIE	4.43	.53
CRS	4.00	.63
FENACOOP	2.67	.58

★ Scale: 1 = strongly disagree; 2 = disagree; 3 = undecided; 4 = agree; 5 = strongly agree. Means are statistically significantly different at a 1% level.

Source: Own data collection and analysis

Table 2.6 Appreciation of the success of different NLA members

Element	Conduct–trust	
Statement	13. The NLA is known to be successful at the things it tries to do.★	
NLA-member	Mean	Standard deviation
FUNICA	4.57	.51
CATIE	4.29	.49
CRS	4.18	.60
FENACOOP	3.33	.58

★ Scale: 1 = strongly disagree; 2 = disagree; 3 = undecided; 4 = agree; 5 = strongly agree.
Means are statistically significantly different at a 5% level.

Source: Own data collection and analysis

developed a more trusting relationship with their NLA counterpart than cooperatives further down in the network at village and community levels. This positive impact of the position of the organization within the network on trust could be explained by the higher frequency of meetings with the project manager from an NLA member for the national and regional cooperatives, in line with similar results on inter-personal trust in the literature on marketing business relationships (Laeequddin *et al.*, 2010). The negative sign of the regression coefficients for the variables related to source of funding (where NGO funding is always the base for the scale) also confirmed that NGOs helping their network partners with funding were more likely to develop trust from their partners.

A second regression model (Table 2.8) showed that, for the cooperative representatives interviewed, factors representing 'trustful relationships' and 'trustful contracts' both had a positive impact on the factor representing 'innovation'. This provides further empirical backing of the importance of trust within the NLA network to reach one of its learning outcomes: improved innovation capacity in agribusiness. However, the lack of statistical significance of the variable 'Connection with NLA' in both regression models also confirmed that the NLA had not had a significant impact on developing trust or improving agribusiness skills of farmers' cooperatives compared with other learning mechanisms in the Nicaraguan agricultural sector.

Suggestions for improvements of the NLA learning cycles

One technician from a governmental institution (who asked to remain anonymous) said that the NLA training guides and their content were very good. However, he mentioned that the way they were taught to farmers was not very successful: the language of the NLA guides was not adjusted to the regional dialect, thus making the training less relevant. Furthermore, the

Box 2.2 Coffee producers' cooperative learns how to manage its books and reputation from NLA partners

In 2006 29 smallholder coffee producers from a Jinotega community formed the '19 de Julio' cooperative to commemorate the Nicaraguan independence date. At first, the cooperative was disorganized: members lacked knowledge on fundamental management processes. Worse, lack of trust between cooperative members and managers, and other economic and social problems, contributed to worsening the disorganization. Having realized the magnitude of its organizational problems, the cooperative was invited to join the NLA learning cycles by CATIE, the national-level research centre. This had a major influence in optimizing their strategic planning and reorganization. CATIE and the second-level agricultural cooperative Union of Jinotega Agricultural Cooperatives (UCA SOPPEXCCA) were both involved in providing training to the representatives of the primary cooperative, with the main goal of improving the living standards of farming families. UCA SOPPEXCCA also supported the 19 de Julio cooperative and its individual members with financial and technical help to strengthen its development. The training provided led to major changes in the cooperative's practices in coffee production and commercialization, enterprise organization, strategic orientation, communication and administration, and dealing with social and environmental issues.

Oscar Antonio Guzmán, a member of the cooperative's executive board remembers: 'Recently, we have been privileged to be trained; we have learned how to produce better on our farms and how we should manage the cooperative better. Because beforehand, we did not know how to manage the register books and now we are doing this by ourselves.'

As a result of the NLA's training, better management has increased members' trust in the cooperative process. They are now able to sell their goods to the international coffee market and the membership has increased to 37 individual members. Ada Lila Lumbi, a female member since 2007, reflects: 'I obtained my land plot through a credit from SOPPEXCCA. From there to now I've seen changes in my life: I've obtained a bit more income. My family has four boys and whatever problem that I have, I consult my cooperative.'

Adapted from Lorio *et al.* (2010, pp. 21–24).

Table 2.7 Regression analysis of selected structure indicators on the factor 'trustful relationships'

Dependent variable: Factor: trustful relationships	Unstandardized coefficients		Standardized coefficients	t	Sig.	Collinearity statistics	
	B	Std. Error	Beta			Tolerance	VIF
(Constant)	.293	.990		.296	.768		
Level of education★[a]	-.302	.123	-.281	-2.464	.016	.587	1.702
Years working for the organization[b]	.025	.014	.162	1.752	.084	.891	1.123
Percentage of male producers that are members of your organization or influenced by it★	.015	.005	.288	2.919	.005	.783	1.278
Position of the organization inside the network★[c]	-.197	.088	-.260	-2.230	.029	.564	1.774
Connection with NLA[d]	-.279	.211	-.138	-1.321	.191	.699	1.430
Did you ever leave a group/IP/cooperative?[d]	-.349	.216	-.160	-1.612	.112	.780	1.282
Active as a producer★[d]	.824	.384	.294	2.146	.036	.407	2.460
Active as a trader★[d]	-.689	.337	-.273	-2.047	.045	.428	2.335
Active as a funding agency★[d]	1.411	.665	.212	2.123	.037	.768	1.303
Active as a financial organization★[d]	.668	.246	.314	2.710	.009	.568	1.761
The most important source of funding is operation generated cash★[d]	-.525	.238	-.235	-2.204	.031	.675	1.482
The most important source of funding is the government[d]	-.579	.429	-.135	-1.349	.182	.764	1.309
The most important source of funding is membership fees★[d]	-.908	.316	-.290	-2.870	.005	.748	1.337
The most important source of funding is credits by the private sector[d]	-.418	.300	-.139	-1.396	.167	.768	1.302
Have you ever shared business/production information with others?[d]	.687	.405	.174	1.698	.094	.724	1.381
The most important channel of communication is the mobile phone[d]	-.839	.465	-.398	-1.805	.076	.157	6.376
The most important channel of communication is the computer[d]	.139	.469	.066	.296	.768	.152	6.575
The most important channel of communication is meetings[d]	-.174	.478	-.074	-.363	.717	.183	5.467

★ Variables with significant influence on the Factor: Trustful relationships.

R-Square = 0.488; Adjusted R-Square = .350; Significance = 0.000; level of significance $p < 0.05$.

[a] Scale: 1 = Primary; 2 = Secondary; 3 = Technical certification; 4 = University; 5 = Postgrade; 6 = PhD. [b] Scale: Years in numbers. [c] Scale: 1 = National organization; 2 = Regional organization; 3 = Cooperative 3rd level; 4 = Cooperative 2nd level; 5 = Cooperative 1st level. [d] Scale: 0 = No; 1 = Yes.

Source: Own data collection and analysis

Table 2.8 Regression analysis of selected structure and conduct indicators on the factor 'innovation'

Dependent variable: Factor: innovation	Unstandardized coefficients		Standardized coefficients	t	Sig.	Collinearity statistics	
	B	Std. Error	Beta			Tolerance	VIF
(Constant)	-1.709	.907		-1.883	.064		
Years working for the organization[*][a]	.044	.013	.294	3.381	.001	.914	1.094
Connection with NLA[b]	.249	.177	.124	1.405	.164	.885	1.129
Position of the organization inside the network[*][c]	-.131	.065	-.178	-2.010	.048	.883	1.132
1. We usually share information about production with other stakeholders[d]	.172	.117	.130	1.467	.147	.881	1.135
11. The NLA/our organization exchange information about their ongoing activities with us[d]	.208	.123	.167	1.690	.095	.711	1.407
13. We plan our activities together with the NLA/our organization according to our production potential and customer demand[*][d]	-.260	.115	-.224	-2.265	.026	.707	1.415
14. Our viewpoints are taken into account by the NLA/our organization when they plan their activities[d]	.028	.142	.022	.201	.842	.558	1.791
15. Joint planning of activities with the NLA/ our organization has improved in the last six years[*][d]	.447	.126	.378	3.541	.001	.607	1.646
10. We prefer to have long-term relationships[d]	-.174	.125	-.127	-1.387	.169	.828	1.208
Factor: Trustful relationships[*]	.252	.096	.248	2.613	.011	.771	1.298
Factor: Trustful contracts[*]	.230	.091	.231	2.532	.013	.834	1.200

[*] Variables with significant influence on the Factor: Innovation.
R–Square = 0.480; Adjusted R–Square = .404; Significance = 0.000; level of significance p < 0.05.
[a] Scale: Years in numbers. [b] Scale: 0 = No; 1 = Yes. [c] Scale: 1 = National organization; 2 = Regional organization; 3 = Cooperative 3rd level; 4 = Cooperative 2nd level; 5 = Cooperative 1st level. [d] Scale: 1 = strongly disagree; 2 = disagree; 3 = undecided; 4 = agree; 5 = strongly agree.

Source: Own data collection and analysis

contents of the guides were applicable to the whole country of Nicaragua, not necessarily accommodating regional niche products. Cooperative representatives and farmers confirmed this latter statement. Aware of these problems, FUNICA has already modified its training guides to address these criticisms.

Some farmers and cooperatives declared that they would like to share information and experiences with each other using the NLA's learning methods but within the same level of the network rather than receiving training from, and extending training to organizations from the network's upper and lower levels respectively. Sharing experiences among farmers' cooperatives at the same level within the network would optimize the method and increase the benefits for the potential participants in this dialogue. Likewise, some cooperatives at first and second levels would like to participate in smaller platforms to improve their performance.

Although the NLA was supposed to be open to the public and private sectors (Lorio *et al.*, 2010), its members currently only consisted of NGOs or research organizations with a similar status in the Nicaragua agricultural development sector. Including representatives of the government- and private-sector-sponsored agribusiness learning programmes within the NLA would help it increase its coverage and incorporate successful learning processes already tested in other national IPs. This would also make the NLA fit better the definition of an IP (Homann-Kee Tui, 2013): a space for different types of stakeholders to get together to solve common problems.

Another criticism of the NLA was that its final beneficiaries were not really those who defined the platform's main goals and methods for achieving them. Indeed, the NLA was part of a bigger platform, the regional learning alliance, where the main goals were set by international development partners and all the participating national learning alliances such as the NLA. The NLA thus used a downstream structure for training where final beneficiaries had little say in what they were going to be taught.

Finally, the NLA's future was still subject to obtaining external funding, as mentioned cursorily in the NLA's strategic planning document (AdA Nicaragua, 2012). Each learning cycle depended on the NLA's donors and how much financial support each NLA member was offering. FENACOOP, for example, had to change their financial planning mid-cycle and the NLA project manager inside FENACOOP left. The fact that FENACOOP stopped working with the NLA because of a funding decision was the reason why FENACOOP was not rated as positively as other NLA members by its partners in the field.

The NLA has already started responding to the feedback it has gathered through its evaluation process and is now addressing all these criticisms. The 2013–2016 strategic plan called for the alliance to adapt better to the needs of the farmers. Furthermore, smaller regional platforms were being fostered and should get more responsibility to tackle the needs of the farmers that are uniquely specific to the different regions of Nicaragua. The NLA also wanted to strengthen its financial situation and develop guides for financial issues at the production level (AdA Nicaragua, 2012).

Lessons learned for other IPs

Although the NLA was not significantly different from other Nicaraguan development networks in achieving positive results in agribusiness skills developed, overall, the levels of agribusiness skills have been increasing in Nicaragua thanks to all the available training initiatives. All these networks have benefitted from the strong cooperative structure in Nicaraguan agriculture and its long tradition of technical training to cooperative members down to the individual farmers. The NLA, governmental organizations, the private sector and other development networks were making the most of this situation to streamline their innovation processes through the cooperative network. Other IPs active in countries with similarly strong networks reaching down to individual farmers should tap into them to foster innovation rather than creating redundant parallel networks.

This study has also showed the importance of the personal involvement of a project manager designated by the NLA member to take part in physical meetings with other NLA members and their target audience in the network. The further away the target audience from the source of learning, the lower the perception of the usefulness of the learning mechanism in building skills. Other IPs should take note of this finding emphasizing the role of a dedicated physical IP facilitator to create a trustful environment between platform members, which will be conducive to information shared and innovations fostered.

Acknowledgements

This work was undertaken as part of the CGIAR Research Program on Policies, Institutions, and Markets (PIM) led by the International Food Policy Research Institute (IFPRI). Funding and support for this study was provided by the CGIAR Research Program on Humidtropics and the CGIAR Research Program on Policies, Institutions, and Markets. This chapter has gone through the standard peer-review procedure of the International Livestock Research Institute. The opinions expressed here belong to the authors, and do not necessarily reflect those of PIM, IFPRI or CGIAR.

Appendices

Appendix 2.1 Characteristics of the farmers' organizations interviewed and their representatives

Individual suervey respondents

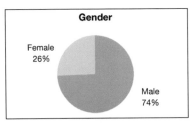

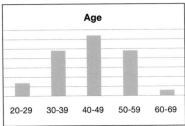

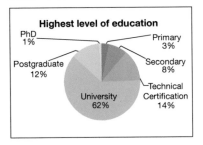

Farmer's Organizations surveyed

Position and connection with the NLA of the organizations surveyed

Position of the organization within the network	Connection with NLA		Total
	Not a member/ No connection	Member/ Connection	
National organization	11	1	12
Regional organization	3	3	6
Third-level cooperative	1	1	2
Second-level cooperative	7	7	14
First-level cooperative	28	26	54
Not applicable/other	2	0	2
Total	52	38	90

Main agricultural product produced by organizations surveyed

	Percent
Coffee	46
Grains (beans, maize, rice)	37
Others (cattle, milk, vegetable, honey, cocoa)	18
Total	100

Main source of funding of organizations surveyed

	Percent
Operation generated cash	27.8
NGO funded	41.1
Government funded	7.8
Membership fees	11.1
Credit (private sector)	12.2
Total	100

Source: Own data collection and analysis

Appendix 2.2 Comparison of data collected between members and non-members of the NLA network

	Membership of NLA network					
	Non-member		Member		Total	
	Mean	Std. dev.★	Mean	Std. dev.★	Mean	Std. dev.★
Age in October 2014	45	10	43	9	44	10
1 We usually share information about production with other stakeholders	4.33	.73	4.21	.78	4.28	.75
2 The information we get from the other business–partners is useful	4.58	.50	4.47	.65	4.53	.56
3 The information we get from the other business–partners/value chain partners is reliable/useful	4.35	.56	4.37	.59	4.36	.57
4 We attend periodic meetings of stakeholders to discuss common production/business problems	4.21	.89	4.39	.72	4.29	.82
5 We use contacts with other actors of the value chain to get information relevant to our business activities	4.35	.71	4.16	.79	4.27	.75
6 We are satisfied with the communication frequency we have with other stakeholders involved in production/business activities	3.81	1.01	3.87	.88	3.83	.95
7 We can express our views freely in exchanges with our value chain partners	4.48	.80	4.63	.54	4.54	.71
8 Our trust on products provided by value chain partners has increased	3.88	1.13	4.05	.73	3.96	.98
9 We have greater trust in our supplier/customer if they are also part of a group we are part of	3.90	1.01	4.05	.90	3.97	.97
10 We exchange information with our value chain partners about our ongoing activities	4.38	.66	4.29	.73	4.34	.69
11 Our value chain partners exchange information about their ongoing activities with us	4.02	.87	4.08	.82	4.04	.85
12 We plan our activities according to the activities of our value chain partners	3.96	1.03	3.74	1.16	3.87	1.08
13 We plan our activities together with our value chain partners according to our production potential and customer demand	3.98	.83	3.89	.95	3.94	.88
14 Our viewpoints are taken into account by our value chain partners when they plan their activities	4.08	.93	4.16	.75	4.11	.85
15 Joint planning of activities with our value chain partners has improved recently	4.08	1.06	4.13	.62	4.10	.90
If yes, how often per year	40.48	102.62	19.71	59.75	31.08	86.08

| | Membership of NLA network | | | | | |
| | Non-member | | Member | | Total | |
	Mean	Std. dev. ★	Mean	Std. dev. ★	Mean	Std. dev. ★
16 Trust is important for the activities with our business partners	4.71	.50	4.71	.52	4.71	.50
17 Our organization's business partners always give us correct information	4.13	.86	4.29	.61	4.20	.77
18 Our organization's business partners always try to inform us if a problem occurs	4.25	.76	4.16	.82	4.21	.79
19 Our organization's business partners always keep their promises	3.75	.81	3.68	.77	3.72	.79
20 The business partners' actions and behaviours are not very consistent	2.46	1.06	2.32	.84	2.40	.97
21 The frequency of contact has a positive influence on the trust	4.31	.76	4.45	.55	4.37	.68
22 Our organization has confidence in all its business partners	3.88	.63	4.03	.82	3.94	.72
23 We only maintain relationships with our business partners with clearly written terms and conditions	3.60	1.25	3.82	1.16	3.69	1.21
24 We only develop relationships with business partners who are fair to us	3.82	.93	4.26	.89	4.01	.94
25 We prefer to have long-term relationships	4.50	.64	4.47	.83	4.49	.72
26 We believe the information provided to us by the groups we participate in	4.19	.69	4.21	.58	4.20	.64
27 The NLA has a lot of knowledge about the work that needs to be done	4.34	.71	4.47	.69	4.41	.70
28 The NLA is known to be successful at the things it tries to do	4.32	.70	4.32	.62	4.32	.66
29 We do not mind paying the NLA subscription fee to get services relevant to us	4.06	.83	3.82	.80	3.95	.82
30 Representatives of the NLA facilitate innovation at the national level	3.72	1.16	3.73	1.04	3.73	1.10
31 Platform members communicate their achievement in other organized groups	3.73	1.11	4.11	.65	3.91	.93
32 The Learning Alliance lobbies for policy changes on national level	3.61	1.23	3.42	1.08	3.51	1.15
33 In the past 5 years, we have applied new techniques or machinery into our production, production process or management	3.81	1.12	3.87	1.14	3.83	1.12

No.	Statement	Mean	Std. Dev*	Mean	Std. Dev*	Mean	Std. Dev*
34	In the past 5 years, we have gained knowledge and skills applicable in our activities from stakeholders outside NLA	4.33	.86	4.16	.82	4.26	.84
35	In the past 5 years, we have gained knowledge and skills applicable in my activities from NLA stakeholders	3.37	1.51	4.39	.59	3.86	1.27
36	We have improved our product in the last 5 years	4.38	.72	4.45	.60	4.41	.67
37	In the past 5 years, there has been an improvement in the interaction between policies, government and other stakeholders	3.62	1.12	3.24	1.30	3.46	1.21
38	We have a better access to the market than 5 years before	4.00	1.02	4.03	.94	4.01	.98
39	The NLA has created smaller platforms at regional/provincial level	3.74	1.05	3.76	.91	3.75	.98
40	The NLA actively supports the work of other IPs at provincial/regional level	3.88	.99	3.89	.95	3.89	.96
41	The NLA encourages us to form working groups within the platform to discuss specific problems	3.87	1.11	3.97	1.05	3.93	1.07
42	In the past 5 years, we have had enough capital for doing new investments	2.87	1.17	2.71	1.04	2.80	1.11
43	It was easier in the last 5 years to get inputs and services needed for our business	3.50	1.08	3.63	.88	3.56	1.00
44	I can get inputs and services at better conditions than 5 years ago	3.69	.98	3.68	1.12	3.69	1.03
45	Total quantity of produced goods has increased since 5 years ago	3.71	1.09	3.89	1.06	3.79	1.08
46	We have developed new products in the last 5 years	3.59	1.31	3.84	1.00	3.70	1.19
47	We have added other activities to our business in the past 5 years	3.83	1.20	4.18	.69	3.98	1.03
48	We have started new cooperation's and joint actions with other business partners in the last 5 years	3.85	1.02	4.03	.82	3.92	.94
49	In the past 5 years, we have adopted new practices in business/production	4.04	.97	4.16	.68	4.09	.86
50	Annual income from business activities has been increasing in the past 5 years	3.22	1.15	3.34	1.10	3.27	1.12
51	We have changed to or entered another value chain in the last 5 years	3.40	1.32	3.61	1.13	3.49	1.24
52	Our networking activities are contributing to some policy changes in government offices	3.55	1.15	3.53	1.03	3.54	1.10
53	Our knowledge about our activity has improved in the past 5 years	4.31	.64	4.42	.60	4.36	.62

* Std. Dev. means Standard Deviation.

Source: Own data collection and analysis

Notes

1   The Learning Alliance started its work in 2003 in four Latin American countries. The initial partners were CIAT, CARE, CRS, GIZ, UNA, SNV, SwissContact and IDRC. IDRC provided financial support. CATIE and VECO Mesoamerica joined the LA later. CRS and CIAT also initiated learning alliances with a similar structure in Africa and in Southeast Asia.
2   However, our model does not use the indicators of the original SCP model because they are not relevant to complex multi-stakeholder innovation systems.

References

AdA. 2014a. 'Alianza de Aprendizaje'. www.alianzasdeaprendizaje.org/portal/index.php. Accessed 21 November 2014.

——— . 2014b. 'Alianza de Aprendizaje-Metologias: Ciclos de Aprendizaje'. www.alianzas deaprendizaje.org/portal/metodologia/24-ciclos-de-aprendizaje. Accessed 21 November 2014.

AdA Nicaragua. 2012. 'Alianza de Aprendizaje Nicaragua: Plan estratégico 2013–2016'. *AdA Nicaragua- Alianza de Aprendizaje Nicaragua.* Managua.

Cadilhon, Jean-Joseph. 2013. 'A conceptual framework to evaluate the impact of innovation platforms on agrifood value chains development'. Paper presented at the 138th seminar of the European Association of Agricultural Economists on Pro-Poor Innovations in Food Supply Chains, Ghent, Belgium, 11–13 September 2013. https://cgspace. cgiar.org/handle/10568/33710. Accessed 17 July 2014.

CATIE. 2008. 'Innovation for sustainable value chains in Central America: Multi-sector strategies to strengthen the capacity of farm households and grower organizations to innovate under ecological and economic uncertainty'. Project report. CATIE, Turrialba, Costa Rica.

FAOSTAT. 2014. 'Nicaragua'. http://faostat.fao.org/CountryProfiles/Country_Profile/ Direct.aspx?lang=en&area=157. Accessed 12 February 2014.

Homann-Kee Tui, Sabine, Adewale Adekunle, Mark Lundy, Josephine Tucker, Eliud Birachi, Marc Schut, Laurens Klerkx, Peter Ballantyne, Alan Duncan, Jo Cadilhon and Paul Mundy et al. 2013. 'What are innovation platforms?' *Innovation Platforms Practice Brief* 1. https://cgspace.cgiar.org/handle/10568/34157. Accessed 21 April 2014.

INTA. 2011. 'Guía metodológica de esculas de campo para facilitadores y facilitadoras en el proceso de extensión agropecuaria'. *INTA-Instituto Nicaragüense de Tecnología Agropecuaria.* www.inta.gob.ni/biblioteca/images/pdf/guias/GUIA%20DE%20ESCUELA %20DE%20CAMPO%20DE%20AGRICULTURA%20ECA%20EN%20EL%20 PROCESI%20DE%20EXT%20AGRI%20FINAL.pdf. Accessed 7 December 2014.

Laeequddin, Mohammed, B.S. Sahay, Vinita Sahay and K. Abdul Waheed. 2010. 'Measuring trust in supply chain partners relationships'. *Measuring Business Excellence* 14 (3): 53–69.

Lafortezza, Daniela and Etimos S.C. Consorzio. 2009. *BCIE 2009 Nicaragua Inventario de las cooperativas productivas.* www.bcie.org/uploaded/content/category/1452167502.pdf. Accessed 2 December 2014.

Landmann, Dirk Hauke. 2015. 'The influence of trust in the Nicaraguan Learning Alliance on capacity development of members and other influenced groups'. MSc thesis. Gottingen, Germany: Georg-August-Universität Göttingen. https://cgspace.cgiar.org/ handle/10568/56689. Accessed 11 February 2015.

Lorio, Margarita, Maria Veronica Gottret and Liana Santamaría. 2010. *Cosechando los Frutos del Cambio Organizacional: 23 organizaciones que con esfuerzo y compromiso trabajan para mejorar el nivel de vida de sus comunidades:* Centro Agronómico Tropical de Investigación y Enseñanza (CATIE). www.alianzasdeaprendizaje.org/portal/documentos/category/28-2-gestion-de-cadenas-de-valor-incluyentes-y-sotenibles?download=260:fase-2. Accessed 7 December 2014.

Lundy, Mark and María Verónica Gottret. 2005. 'Learning alliances: An approach for building multi-stakeholder innovation systems'. www.cgiar-ilac.org/files/ILAC_Brief 08_alliances_1.pdf. Accessed 13 May 2014.

Ruben, Ruerd and Zvi Lerman. 2005. 'Why Nicaraguan peasants stay in agricultural production cooperatives'. *Revista Europea de Estudios Latinoamericanos y del Caribe 78:* 31–47.

The World Bank Group. 2012. 'The International Development Association and International Finance Corporation: Country Partnership Strategy (FY2013–2017) For The Republic of Nicaragua'. (Report No: 69231-NI). www-wds.worldbank.org/external/default/WDSContentServer/WDSP/IB/2012/10/24/000386194_20121024011712/Rendered/PDF/692310CAS0P1280Official0Use0Only090.pdf. Accessed 2 December 2014.

—— . 2014a. 'Nicaragua'. www.worldbank.org/en/country/nicaragua. Accessed 2 December 2014.

—— . 2014b. 'Nicaragua: Nicaragua Overview'. www.worldbank.org/en/country/nicaragua/overview#1. Accessed 2 December 2014.

3 Overcoming challenges for crops, people and policies in Central Africa

The story of CIALCA stakeholder engagement

Perez Muchunguzi, Piet van Asten, Bernard Vanlauwe and Guy Blomme

Intercropping of banana and coffee is not allowed officially, is this going to change? Research has shown promising results. I have 1 ha of mixed banana and young coffee, now I have to choose only one crop because of official recommendations. So can I keep both crops? Can I also go and tell other farmers to intercrop their banana with coffee?

(A male farmer from Musaza sector, Kirehe district asking policy makers during a stakeholders meeting organized by the Minister of Agriculture of Rwanda)

Introduction

The great lakes region of Central Africa is beautiful and abundant in hills, people and conflicts. Its high altitude and cooler climate make it ideal for crops. But soils have been exhausted, spare land is rarely available, and competition and struggle for resources has marked much of the region's history of the past 50 years. Many farmers in parts of this region rank among the most food insecure and malnourished on earth. This is because of low farm productivity since the majority depends on agriculture that is done with minimal fertilizer use. A 2006 baseline survey revealed that more than 60 percent of the population in Central Burundi and South Kivu were food insecure and had very few opportunities to diversify income with off-farm activities. Farm sizes are too small (< 2 ha). Although Democratic Republic of Congo (DRC) still has some spare land, the existing land tenure arrangements do not encourage farmers to invest in soil and water conservation since most of the land is in the hands of the chiefs, locally known as "Mwamis." These challenges, nested across different scales, point to the need for innovative ways of working through multi-stakeholder processes.

The Consortium for Improving Agriculture-based Livelihoods in Central Africa (CIALCA) was set up to provide science-based evidence that helps bridge the knowledge gap between farmers, public and private extension workers,

scientists and policy makers. Thousands of farmers usually find themselves in a dilemma similar to that of the male coffee farmer quoted above—wanting to respond to their practical challenges on the ground but finding themselves constrained due to non-matching policies or institutional settings. On the other hand, policy makers also lack credible evidence on which to base their decisions. In this farmer's situation, planting coffee ensures a seasonal harvest of cash. But he also wants to be food secure from the same piece of land and so planting bananas in his newly planted coffee makes sense since, in addition to food, bananas will provide a steady cash flow throughout the year. Coffee is a big foreign exchange earner for the country and so farmers are discouraged from intercropping the two crops. Due to realization of the land shortage though, farmers are sometimes allowed to plant bananas in the coffee when the coffee is still very young. They are however required to cut the bananas when the coffee has reached its productive stage. Greater in number even are farmers restricted by knowledge and resources, not policies. The combination of these factors made CIALCA realize that registering any meaningful changes required many more stakeholders at the table ranging from farmers to policy makers, and this was how the CIALCA work shaped into "platforms" at different levels to serve different but connected needs.

Emergent IPs

CIALCA started out as an inclusive research consortium for development, spearheaded by three international agricultural research centers: The International Institute of Tropical Agriculture IITA, Bioversity International and the International Center for Tropical Agriculture, CIAT. It started in 2005, but was formally launched in 2006, operating in the three countries (Rwanda, Burundi and DRC). Each country had its own challenges as well as opportunities and this called for different methods of engagement in each of the countries. In Rwanda for example, where strong national policy shapes smallholder farming, the Consortium developed platforms around the government's research and extension systems. In the DRC and Burundi that were still recovering from conflicts, the scaling/extension component was handled through the non-governmental organizations (NGOs) but research components were still handled through the National Agricultural Research Systems (NARS).

Different types of "innovation platforms" emerged across different levels. These brought together different stakeholders operating in these geographical sites. The levels included field sites in a village or local community (usually around an experimental field), and an action site that was equivalent to say a district, national and regional level between countries. The coming together of the stakeholders fostered cross-learning and experience sharing. The learning was usually organized through the field days at the field sites, and through meetings and conferences at the action site and other levels (see Figure 3.1).

Figure 3.1 Farmer field day bringing stakeholders together (left) and CIALCA conference organized in Kigali 2011 (right)

Photos: CIALCA

Bernard Vanlauwe, CIALCA Scientist says:

> At that point in time we had not heard much about innovation platforms. These platforms simply emerged out of need, which was key for their crucial role in fostering adaptive collaboration between different groups of stakeholders, and CIALCA's impact and reputation in the region.

The platforms emerged and grew as the need arose. For example CIALCA's collaborative work to fight Banana Xanthomonas Wilt (BXW) with the Rwandan government research and extension departments and the regional stakeholders in the Rubavu area clearly required multiple stakeholders. The "simply emerging" nature of these platforms helped to avoid many expectations and allowed an organic means of platform evolvement. Inclusion within the platform was based on mutual needs fulfillment rather than position filling. This evolvement was very befitting as the CIALCA team was extensively made up of natural scientists that would have found difficulties in managing the different stakeholder expectations.

Embracing the work challenge

Following a series of 25 participatory rural appraisals across the region, the consortium decided to focus its agronomic interventions on key entry points in smallholder cropping systems; i.e. bananas, (soy)beans, coffee, cassava and maize. These crops are vital sources of food and revenue, yet their productivity is chronically hampered by inferior planting material, crop diseases, poor agronomic practices and limited capacity to access markets as well as restrictive policy environments.

CIALCA proposed to work on these cropping systems to contribute to its overarching goal of improving the livelihoods of those who depended on agriculture through research investments in system productivity and resilience.

Research for development activities were varied and dynamic, but greater emphasis was placed on introducing and evaluating better banana and legume germplasm, improving agronomic practices in mixed cropping systems, integrated soil fertility management, integrated pest management, and social innovations for improved crop marketing leading to income. CIALCA made investments in developing intercropping options for staples such as banana and cassava with legumes. Most of these technologies were already being practiced by farmers elsewhere in the East African highlands with varying degrees of success: (i) banana–coffee intercropping concepts were transferred from Uganda to Burundi and (from there) later to Rwanda, (ii) zero-tillage mulch banana x bean intercropping was transferred from Uganda to Burundi and Rwanda, (iii) smart legume intercropping systems in maize and cassava were first tested in West Kenya and subsequently successfully tested and documented in DRC, (iv) soybean processing technologies moved from Uganda/Kenya to Rwanda and DRC. Several key recommendations were made based on this research that is being out scaled, for example the using of sticks to make holes in the banana mulched plantations demonstrated below, ensured minimum soil disturbance providing the much needed source proteins while keeping the banana root system intact (see Figure 3.2).

These thematic areas responded to partner needs identified from the baseline survey as well as participatory rural appraisals. New varieties of the bananas, cassava and legumes were introduced and jointly evaluated by the stakeholders. This was usually done in field trials that were strategically positioned at a field site in a village or local community. Learning and experience sharing was usually carried out through field days while partners active within an action site usually met in organized meetings.

Figure 3.2 Beans intercropped with bananas benefit from each other
Photo: CIALCA

The banana, coffee and legume intercropping technologies promoted by CIALCA in the region is one way to best demonstrate this. In general, planting bananas with coffee at the right ratios improved labor-use efficiency, overall income by >50 percent and reduced farmer's exposure to climate shocks (van Asten *et al.*, 2011). While the agronomic and economic benefits were clear from the research and farmers side, the institutional policy arrangement to make these benefits available to farmers were nonexistent. This then created a need for policy actor engagements.

System synergies and trade-offs: Coffee–banana integration: win–win–lose?

Despite the coffee–banana intercropping benefits, there was an emerging gender challenge (see Table 3.1). Across the region in general, men often cited a stronger labour investment by women in the management of coffee plots when intercropped with cooking bananas as the women care for the food security of the household. This however brings a strong gender-biased division of farm enterprises, resource control, and task execution, which seems to provide a serious disincentive to really improve resource-use efficiency at the farm level.

Table 3.1 Pros and cons about coffee–banana integration

Pros	Cons
• Increased productivity • Increased income and food security • Better cup quality	• Coffee is largely dominated by men in the region. Intercropping means that men are benefiting from labour coming from women as they attend to the food crop
• Better resilience to market volatility	• The productivity can go down drastically if the banana and coffee densities are not properly managed

Source: van Asten *et al.* 2011

Due to its regional nature and focus on multiple commodities, CIALCA's activities have now been integrated into the CGIAR research program Humidtropics, which aims to help poor farm families in tropical Africa, Asia and the Americas boost their income from integrated agricultural systems' intensification while preserving their land for future generations.

Understanding and exploiting diversity at the farm level

The consortium stakeholders mapped the flow of resources and quantified soil fertility gradients and on-farm nutrient recycling across sites. Our results showed that farmers disproportionally favor home-gardens in terms of nutrient and labor inputs, often relying on perennial crops and vegetables in homestead plots that are more fertile. Our quantification of the nutrient stocks and recycling showed that it was absolutely vital to keep crop residues on farm, since this would reduce nutrient losses for many crops by 50 percent or more.

Consequently, given the importance of erosion in the hilly region, the researchers and their local partners conducted a number of integrated technology trials to try to improve productivity while reducing erosion. Technologies tested in various combinations were (i) embankments, (ii) hedge-cropping, (iii) no-tillage. Many were surprised to discover that the various erosion control options did not lead to the aspired improved productivity. On the contrary, all the technologies actually reduced yield of the maize and soybean being cropped together. Just as disappointing, the increased labor, competition for water and space, and soil disturbance to make the embankments did not help to improve productivity over the 1–2 years of the trial. Additionally, the fact that in eastern DRC, the "Mwami" land tenure system did not favor the majority poor farmers growing crops on the land also gave no incentive for farmers to make any meaningful investments in erosion control. The consortium experienced this first hand when one of the experimental field trials was taken away after the landlord had seen that fertilizers had been applied. While this was a loss for experimental data collection, it was by far one of the most natural ways to understand the day to day difficult decisions that the land renting farmers have as a result of the land tenure system.

Innovations delivering impact

From 2006 onwards, socio-technical innovations through platforms sought to improve the livelihoods of poor farmers in Burundi, Rwanda and DRC by enhancing their capacity to improve agricultural productivity for better income, nutrition, and environment. CIALCA demonstrated and disseminated solutions to some of these pressing problems:

- Introduced exotic banana varieties proved extremely popular with farmers and extension partners in certain areas. They are very well adapted to local growing conditions, often yielding double the bunch-weight of local varieties.
- Legume germplasm introduced by CIALCA was rapidly out-scaled through farmer-led seed multiplication in Bas-Congo and the Eastern Province of South Kivu. More than half of the farmers involved in these schemes adopted the improved seed.
- An increased production of soybean has prompted the further development of, and trainings on, various highly nutritious soybean products. These trainings particularly target women, resulting in significant nutritional benefits for the young children in their care.
- An innovative banana–coffee intercropping promises increased farm incomes, and increases the resilience of coffee systems to a warming climate. This has

caught the attention of Rwandan and Burundian authorities, who are actively engaged in validating the technology.

- Xanthomonas wilt of banana steadily conquered a large part of the East African highlands. CIALCA contributed to the fine-tuning of an integrated control and rehabilitation package and collaborated with numerous development partners to mitigate disease impact and halt the spread of the disease into new areas.

Cassava–legume intercrop systems saw significant improvements through the use of fertilizer in combination with manure or compost. Legume and cassava yields have increased by at least 40 percent and 20 percent, respectively.

The Consortium chose three measurable criteria to track progress towards their goal: increasing farm level productivity, improving protein intake and boosting household income. The Consortium anticipated that at the end of the project, 2.1 million people would be aware of CIALCA-related activities of which 400,000 were actively seeking access to knowledge and technologies promoted by CIALCA. They set these milestones at project inception in 2006, and introduced a monitoring process during implementation. Finally, CIALCA evaluated the project at its closure in 2011. In the report (Macharia *et al.*, 2012) the key findings were:

- CIALCA's interventions improved farm productivity. In the intervention areas, a rapid impact assessment showed that CIALCA innovations had increased average farm level productivity by more than 27 percent. Some yields have increased up to 179 percent.
- CIALCA increased protein intake. Averaged across all of the CIALCA intervention areas we have demonstrated that adoption of CIALCA technologies significantly increases protein intake. The consumption of protein has increased by at least 12 percent.
- CIALCA has increased household income. By adopting improved agricultural practices and market-oriented strategies, a rapid impact assessment indicates that aggregate household income has increased by over 19 per cent. In some areas, farmers earn an additional 60 to 90 USD per year from improved banana production and marketing.

Different areas, different institutional collaborative arrangements

CIALCA commissioned a study conducted in 2011 to describe the organization of CIALCA: how it came together, how it has adapted to seek out impacts, and where the model's particular style of partnerships has succeeded or fallen short in the eyes of its participants.

This study (Cox, 2011) noted that the foremost asset of CIALCA's functioning was its adaptability, which has brought successes in some drastically different country contexts: in Rwanda for example, where strong national policy shapes smallholder farming, the Consortium came to work very closely with the

---

Box 3.1 How can a project with few staff achieve impact at a regional level?

CIALCA has made considerable investments in making sure new technologies reach (and are able to be used by) partners and farmers. A knowledge resource center was established in 2010 and works closely with partners to identify "best-bet" impact pathways for technology out-scaling. The center also supports the development and packaging of project knowledge in suitable formats (including radio and video) and languages that clearly communicate the actions required. The training-of-trainers (ToT) approach is a central pillar of CIALCA, ranging from crop production to marketing and nutrition related trainings. CIALCA has organized a total of 159 training events and collaborated with over 60 NGO partners and public extension services for its development-oriented work.

Source: Macharia *et al.* (2012)

Photo: CIAT/N. Palmer

---

government's research and extension system. Through this, policy engagement was done, the partners trained farmers in Integrated Soil Fertility Management (ISFM) and the use of newly subsidized fertilizers, and helped the country manage the menace of BXW. In the study mentioned above, when the partners and CIALCA staff in Rwanda were asked about their perceived advantages of working with CIALCA, the top two reasons cited were stakeholder engagement, especially farmers, as well as capacity building. Interestingly, policy engagement is also cited as a strength that CIALCA enjoyed. In Burundi and DRC, where national systems are weakened by recurring civil conflict, CIALCA collaborated with a whole assortment of governmental and non-governmental agencies in identifying and disseminating improvements to banana- and legume-based systems. In both Burundi and DRC, the top ranked advantages associated with CIALCA as perceived by partners and staff included introduction of new varieties and means of multiplying them. Since public service provision was relatively weaker, working through NGOs whose mandate focuses on input provision and training gave better returns in Burundi and DRC. On the contrary, in Rwanda, focusing on and following the processes sometimes took longer than desired but gave better and sustainable returns. CIALCA developed communication materials that were widely adopted and distributed by the government extension arm.

The same applies to the rapid propagation of bananas for example, which was adopted and used by the government extension system as a means to produce healthy planting materials while in both Burundi and DRC this was extensively done by the NGOs.

Capacity development of the actors

From three autonomous regional offices, CIALCA connected with dozens of civil society organizations and NGOs, and community-based organizations (CBOs). These were trained in different technology packages and through Training of Trainers (ToT) across the region. They in many cases were responsible for reaching areas where CIALCA was not working. In north Kivu Eastern DRC, the radio program that one of the CIALCA staff conducted was found to be effective, especially in relaying messages on the control of the BXW (Figure 3.3).

Furthermore, since its inception, CIALCA was strongly committed to capacity building in a region that had lost much of its best agricultural researchers during the long period of conflict and strife. CIALCA has trained over 20 PhDs, 35 Masters and over 135 Bachelor of Science students who now

Figure 3.3 Farmers using CIALCA communication material during the agricultural show in Rwanda

Photo: CIALCA

occupy strategic jobs such as Directors and Department Heads in national research institutes, central governments and beyond.

Finding the link between good science, stakeholder engagement and impact—the role of partnerships

CIALCA used the different regional experience and scientific evidence coming from trials and surveys to engage the different stakeholders from farmers and extension workers to policy makers in order to influence policy changes. This was not completely familiar ground for CIALCA because we learned that knowing people that know other people helps if you can exercise patience to wait for a policy maker for four hours and have a ten minute discussion. For example, in Rwanda, where the government had virtually adopted a policy of sole cropping to encourage farmers to seriously invest in improving crop production following "green revolution" principles, providing the evidence for intercropping proved vital. Farmers did not always agree with this approach since they wanted to earn money but also be food secure on their small pieces of land. CIALCA research and policy actor engagements on the benefits of intercropping systems managed to provoke some reflection at the national policy level. For example, results on the benefits of banana–coffee intercropping (including improved climate adaptation and cup taste) led to the Minister organizing a meeting with all key public actors, NGOs and farmer representatives to discuss the results. These results generated a lot of debate from the different stakeholders ranging from farmers and researchers, as well as extension workers. This was made possible because of the regional platform sharing results between countries. The point was further proved by the Ministry of Agriculture (MINAGRI) website: "the idea of coffee–banana intercropping was first introduced by (assistant agronomist sic) Dr. Van Asten Piet about two years ago. Since then there have been several studies and analyses and lessons learned from Uganda, Burundi and Rwanda itself." This further shows that the CIALCA regional platform was recognized in each of the countries.

While writing about the one-day engagement between CIALCA and the Rwandan Agriculture stakeholders, the MINAGRI website gave a very memorable and potential game changer quote that truly highlighted the role of engaging in multi-stakeholder processes: "This workshop is an indication of a change that may occur within the agriculture sector for Rwanda that will ultimately benefit rural farmers and market prices for the country, as research continues" (Rwandan Ministry of Agriculture, n.d.).

To a large extent, the position presented by the ministry website strongly mirrored the sentiments of the majority of the stakeholders in the workshop. One of these stakeholders represented the Belgium Technical Cooperation, BTC. The BTC representative Mr. Somers Raf said:

> As an extensionist, my question is when to start doing this? The only issue to be confirmed is coffee cup quality. So far, there is no single trial showing

that banana–coffee intercropping affects yields of either of the two crops negatively. Yes, researchers may do their work still for many years, and better density recommendations may be developed after more experiments are done. However, we need to start. After cup quality is confirmed, the only question is why farmers may not start doing it immediately?

Cup quality results of coffee intercropped with bananas later came out and there was positive correlation between intercropped (shaded) coffee and cup quality, further proving the fact that the shade from the bananas had a positive effect on coffee quality. This engagement led to a shift in policy discourse from the key decision makers in the sector—governments no longer consider banana intercropping as a "crime" and in several regions they are actively encouraging intercropping through government-supported farmer field schools. The national research and extension arm of government, the Rwanda Agricultural Board, RAB, has picked the banana–coffee intercropping system and demonstration fields are being set up. This is a real shift in the institutional environment for smallholders who were previously "punished" for intercropping in banana or coffee fields. As noted by MINAGRI, this change has further opened their interest in developing intensified and well-organized intercropping systems that they would like to promote to smallholder farmers.

This had a big impact and implications as it came towards the end of CIALCA. The resources that had been invested in the banana–coffee research, the long-term engagement with the ministry and other key stakeholders and the affirming voices that were heard during the discussions, all pointed to how the process and the content need to work together to have meaningful outcomes. At several critical stages, when gray areas emerged causing tension between stakeholders, the engagement process benefitted from scientific evidence for moving forward.

Learning from the past, looking to the future

Within the CGIAR, the CIALCA consortium was an absolutely unique collaboration when it started in 2005, both in terms of systems approach, as well as in its philosophy of equal partnership and adaptive management.

A number of factors can be pointed to when it comes to what led to the success of the CIALCA platforms:

* *having an evidence-based engagement process*: the research that was done by CIALCA stakeholders led by the NARS in the different countries gave very interesting and new insights that benefitted the engagement process with stakeholders across levels including farmers, civil society and policy makers. This evidence from "good science" kept the partners engaged even when the process was sometimes challenging due to the fact that CIALCA's work was covering a very big area in addition to tackling policy related matters;

- *building on existing knowledge-learning from farmers*: enormous amounts of knowledge already exist within communities. Many of the technological innovations used by CIALCA were based on successful smallholder experiences elsewhere in the East African Highlands and were not necessarily developed from "scratch." Building on this knowledge gave better and quicker place owned results. For example, the coffee–banana system that was studied widely in the CIALCA region was first and foremost picked from practicing farmers especially in Uganda where they provide cash and food security;

- *multiple level engagements/platforms*: platforms engaged at different levels allowed multiple-level exchange of knowledge and expression of needs. From the villages, to field sites, to action sites and to regional (country to country) exchanges. This multi-level organization facilitated site specific as well as between-sites cross-learning. This allowed, for example, policy makers to hear from the farmers in a very organized and effective way that fostered changes. Regional exchange of information was also easy and acceptable as there was recognition of the region as a single block/platform. Information exchange across countries fostered quick and trusted awareness creation. The research generated in one area/country only required validation in the other countries and this saved a lot of time;

- *capacity development*: training of different partners, both formally and informally, did not only improve opportunities for these platforms to handle issues by themselves but also created an opportunity for CIALCA approaches and opportunities to continue in the future in different ways. Capacity building of stakeholders improved the quality of engagement of the stakeholders. For example, it empowered farmers to pose questions to policy makers as long as they knew that they had back-up information. Several graduates have been promoted to senior positions within the Rwanda and Burundi national research systems, attesting to a significant return on investment of research leadership;

- *management and operational flexibility*: the differences between and within countries were too wide to have a "one size fits all" approach. Flexibility in different countries and at different levels allowed a more efficient and cost-effective way to work across countries/levels. Having the flexible donor that walked the journey with CIALCA allowed engagement and implementation to always suit the needs and opportunities within each area without necessarily following the blue print. This was a great incentive for CIALCA's systems work. This was particularly useful as we worked with the multi-stakeholders since the process was in many cases determining the direction.

While progress was made on a number of fronts, the consortium agreed that there were areas that called for improvement. One of those areas identified was that improvement could be made by taking a more holistic approach to its research for development processes: integrate livestock, gender and business planning. The systems learning and policy engagement could also be strengthened further to deal with issues such as land tenure that require much wider social–political engagements that consider factors and approaches beyond land conservation trials.

One question to openly pursue as CIALCA "platforms" move into a formal setting within the CGIAR research program Humidtropics, is how far do we necessarily institutionalize platforms across the region but still allow an organic and adaptive style of operation and management that encourages place-based innovations to freely emerge.

Acknowledgments

Special acknowledgments go to Humidtropics capacity building and all the editors/mentors for the training provided. This helped to shape the case into what it looks like today. The National partners and the farmers in Burundi, Rwanda and DRC that made the CIALCA work a success. We thank you for believing in the work that this partnership produced, thank you very much. The Consortium for Improving Agriculture-based Livelihoods in Central Africa (CIALCA) is a Consortium of the International Institute of Tropical Agriculture (IITA), Bioversity International, and the International Centre for Tropical Agriculture (CIAT) and their national research and development partners, supported by the Belgian Directorate General for Development (DGD), and aiming at improving livelihoods through enhancing income, health and the natural resource base of smallholder farmers in Central Africa.

References

Cox, P.T. 2011. Describing the CIALCA organizational model. *CIALCA Technical Report* no. 16. CIALCA, Bujumbura, Burundi, Bukavu, DRC, and Kigali, Rwanda.

Macharia, I., Garming, H., Ouma, E., Birachi, E., Ekesa, B., de Lange, M., Blomme, G., Van Asten, P., Van-Lauwe, B., Kanyaruguru, J., Lodi-Lama, J., Manvu, J., Bisimwa, C., Katembo, J., Zawadi, S., 2012. Assessing the impact of CIALCA technologies on crop productivity and poverty in the great lakes region of Burundi, the Democratic Republic of Congo (DR Congo) and Rwanda. *CIALCA Impact Evaluation Report.* CIALCA, Bujumbura, Burundi, Bukavu, DRC, and Kigali, Rwanda.

Rwandan Ministry of Agriculture, n.d. www.minagri.gov.rw/index.php?option=com_content&view=article&id=716%3Aminagri-and-rab-participated-in-research-validation-on-banana-coffee-intercropping&catid=154%3Alatest-events&Itemid=270&lang=en (accessed April 5, 2013).

van Asten, P.J.A., Wairegi, L.W.I., Mukasa, D., Uringi, N.O., 2011. Agronomic and economic benefits of coffee–banana intercropping in Uganda's smallholder farming systems. *Agricultural Systems* 104: 326–334.

4 Can an innovation platform succeed as a cooperative society?

The story of Bubaare Innovation Platform Multipurpose Cooperative Society Ltd

*Rebecca Mutebi Kalibwani, Jeniffer Twebaze,
Rick Kamugisha, Honest Tumuheirwe,
Edison Hilman, Moses M. Tenywa and
Sospeter O. Nyamwaro*

Introduction

Linking small scale farmers to markets using value chain approaches has become an important component of many agricultural development interventions in developing countries. Traditional agricultural interventions primarily focus on farm productivity, to ensure food security among households, and their capacity to market the surplus. Agricultural cooperatives often target farmers who are already engaged in growing cash crops. Cooperatives ensure that farmers maintain access to critical farm inputs, market farm products, strengthen farmers' bargaining power and improve income opportunities. However in Uganda there are problems that led to the near-collapse of the cooperative sub-sector. These problems include poor management, and political interference among others (Kwapong and Korugendo, 2010). IPs target a wide range of farmers; those that are still ensuring food security and those already participating in the market. There is therefore need for supporting IPs embracing cooperative societies' approach to avoid similar pitfalls. Such support would entail IPs applying an enhanced cooperative society's model to generate wider and more attractive benefits. This case study therefore shows how the registration of Bubaare IP as a cooperative society has opened market opportunities for its members. It also aims to identify the factors behind this success, the challenges it faces, and draws lessons for wider use by other IPs and the cooperative sub-sector in Uganda.

The case of four IPs in Uganda

The Sub-Saharan Africa Challenge Program (SSA CP) established four IPs in south-western Uganda in 2009. Using the Integrated Agricultural Research for Development (IAR4D) approach, the IPs were established to link small scale farmers to markets by developing commodity value chains, among other things. The success stories of the four IPs after only three years of the pilot phase included increased quality and quantity of production, improved household incomes as a result of linking to markets and getting farmers involved in value addition activities (Adekunle *et al.*, 2013). While IPs have since gained popularity as an approach that promises to lift smallholder farmers out of poverty, little was it known that they would end up as cooperative societies. Four years after its establishment, Bubaare IP in 2013 led the formal registration of the IPs in south-western Uganda as an IP cooperative society in pursuit of market opportunities.

While this registration of the IP was considered to be an achievement among some stakeholders and partners, it was seen as illogical by others, especially given the past performance of the cooperative sub-sector in Uganda. The underlying doubt is whether the success potential driven by actor innovation will surmount the inherent weaknesses in the cooperative sub-sector. The potential is great, because there are efforts to revive the sub-sector through a tripartite cooperative model that integrates the traditional services of savings and credit with marketing through Rural Producer Organizations (RPOs) and Area Cooperative Enterprises (ACEs) (Kwapong, 2013). What is different, besides the ongoing efforts in Uganda is the evidence (from elsewhere in the world) that the IP approach will benefit the cooperative sector. For instance, in India, smallholder farmer groups are registering successful companies. Lessons from this Indian experience can be adapted to support development of IP cooperatives in Uganda, critical to align the sub-sector with the evolving needs of small scale farmers.

The case study is developed from interviews with selected IP coop society leaders, leaders of member Self Help Groups (SHGs), Kabale District Local Government (KDLG) officials, and by analysing IP records and data from partners.

History of Bubaare IP

Background

Bubaare is one of the IPs that was formed by the SSA CP. This is a research programme developed, funded and implemented by the Forum for Agricultural Research in Africa (FARA). The SSA CP, implemented between 2008–2010, employed the IAR4D, an innovation-based research approach involving many stakeholders and innovative partnerships. This approach enabled simultaneous work on most categories of agricultural problems. In the Lake Kivu Pilot

Learning Site (LKPLS) where Bubaare IP is found, IPs were formed around chosen value chains (e.g. sorghum, potatoes, beans) selected jointly by all stakeholders. A total of 12 IPs were formed in the LKPLS, four in each of the three participating countries; Uganda, Rwanda and the Democratic Republic of Congo.

Bubaare IP is found in the Bubaare sub-county of Kabale District in south-west Uganda. Sorghum was initially selected as the enterprise of focus. This is because every household in Bubaare grows a traditional sorghum variety that has been used for generations to produce porridge and weaning food for babies. Sorghum grown traditionally is a socially and culturally important crop but low yields and tedious work have made it unprofitable. Besides, locally processed products, including weaning food for babies only last about three days. Therefore the IP chose to pursue value addition as the key driver of sorghum value-chain development. Led by an executive committee, the IP set out to form a strategy for increasing production and value addition through improved processing and creation of market linkages.

Achievements at the Bubaare IP

Since its establishment in 2009, a number of innovations have been generated in the IP to support the development of the sorghum value-chain and link farmers to the market. These innovations include:

Improved farming practices

One of the challenges identified as key to the development of the sorghum value chain was to raise the quantity and quality of sorghum produced by the farmers before looking for market opportunities. Two local wild varieties of sorghum, *kyatanombe* and *omukoba*, are commonly grown as staple food. They mature in seven months and are harvested once a year. With facilitation by the SSA CP, the members were introduced to improved agronomic practices such as correct plant spacing. Kabale Zonal Agricultural Research and Development Institute (KAZARDI), one of the partner institutions, through field trials with the farmers, developed varieties of sorghum that mature in a relatively shorter time and have higher yields. Currently, about 50 per cent of the IP members are adopting recommended spacing for sorghum, by for instance planting in lines as opposed to broadcasting, and applying fertilizer to improve yields.

Bye-law formulation

When IAR4D facilitators interacted with IP members in October 2009, they realized a crucial need. The community wanted to revise their natural resource management (NRM) bye-laws in order to enhance the development of the sorghum value-chain from production to marketing. The major challenges concerning community bye-laws were not only their poor implementation and enforcement, but also lack of a review mechanism for them to maintain relevance, and to formulate new ones when the need arose. IP members mobilized their respective parishes and villages for the process to review and formulate community bye-laws at the beginning of 2010. The bye-laws entitled Bubaare Sub-county (Natural Resource Management, Agriculture and Marketing) Bye-laws, 2010 were finally approved and signed in a sub-county council session in November 2010. Although with the challenges of enforcement, the bye-laws have been implemented in several parishes, and used to protect gardens and guide marketing procedures.

The sorghum beverage – Mamera

One of the stakeholders that was identified to join the sorghum IP was a food processor, Mr Julius Byamukama, of Huntex Ltd. Located near Kabale town, Huntex is a food processing company, with an outlet supermarket in Kabale town. With Julius as a stakeholder in the IP, it was possible to negotiate with him to process and pack the sorghum produced by the farmers into a beverage. Normally the sorghum is prepared into a drink called Bushera that is used as a beverage in many homes in Kabale. It is also sold in kiosks and shops in cups and simple polythene packing. It may last up to two weeks but the longer it stays, the more alcoholic it gets. After successful negotiations, Huntex Ltd began to process the sorghum purchased from the IP members into a non-alcoholic beverage called Mamera and packed in 500 ml plastic containers (Figure 4.1). This made the drink more hygienic and attractive in presentation. Mamera has a shelf-life of six months and comes in two types; one sweetened with honey and the other without. They each cost UGX1,000/= (USD0.4).[1]

The IP members found it necessary to look into other products to attract a wider market for the sorghum, given the improvements in yield. With the help of the Department of Food Science and Technology at Makerere University, the IP members produced two types of sorghum flour; unmalted sorghum for food and malted sorghum for porridge. These have been packed in 1 kg packets and will soon be launched in supermarkets in Kabale and the rest of the country. Each kilogram packet costs UGX3,500/= (USD1.3). The farmers decided to take on the production of sorghum flour since it is less complicated, while Huntex Ltd produces the beverage.

Other value chains and innovations

After the research phase for the proof of the IAR4D concept, it became necessary to include other enterprises that the farmers on the IP may be involved in, instead of just one as was previously the case. For the Bubaare IP in particular, honey and Irish potato were included. The IP members produce and pack 500 ml jars of honey and sell each at UGX8,000/= (USD2.2). The members have purchased equipment for the production of potato crisps. They are packing the potato crisps in two different sizes; a small one that costs

Figure 4.1 The different packages of Mamera
Photos: Huntex Ltd

UGX1,000/= (USD0.4) and a bigger size that costs UGX2,000/= (USD0.8) (Figure 4.2). These will also soon be launched in the market.

The IP purchased a computer and members are being helped to search for market information on the Internet. The IP members have been introduced to savings mobilization and credit by a partner, Agriculture Innovation System Brokerage Association (AGINSBA). The IP savings are kept with financial institutions, specifically Crane Bank and Muchahi Savings and Credit Cooperative (SACCO), a strong farmers' savings and credit cooperative in the area.

Factors leading to the success of the IP

A number of factors are responsible for the success of Bubaare IP linking its members to the market. The political stability in the country has provided a supportive environment for the private sector to develop. The government has also pursued a conducive macroeconomic policy environment and a decentralized form of governance that supports innovativeness in the way specific localities are able to deal with development challenges. The local government, both at the district and sub-county, have given plenty of support to the IP activities. The sub-county administration has provided the venue where the IP can hold meetings, a room for computer training, a store for the produce of the IP members, and security for the IP property. The stakeholders

Figure 4.2 The different products made by the Bubaare IP members: 1 kg packet of sorghum flour (top left); 500 ml bottle of the Mamera beverage (top right and bottom right); potato crisps (bottom left)

Photos: Bubaare IP

identified to join the IP were relevant to the identified challenges of the community to be addressed by the IP. Huntex Ltd, KAZARDI, KDLG, Ministry of Trade, Tourism and Industry (MTTI), NGOs with good facilitation and networking have all contributed to these developments.

Why the IP formally registered as a Cooperative Society

Bubaare IP was initially registered as an association to be able to operate within the district. It was able to get into all the above ventures and interactions as an association. There was development and a lot of excitement among the IP members, other stakeholders and partners. The IP won special interest from the development partners who gave a grant of USD30,000 in 2012 for the expansion of sorghum production by the IP members, and subsequent processing into larger quantities of Mamera. The IP stakeholders had to decide on the best way to utilize the funds that had been given. The best option turned out to be that the funds be loaned to Huntex Ltd to expand their premises and purchase the required equipment to handle the processing and packaging of larger quantities of the Mamera. The production of sorghum from the fields would be increased through more extension effort and mobilizing more members into the IP.

It was at this point that it necessitated the IP to be formally registered beyond the level of an association. Only then would it be able to sue and to be sued in courts of law in the event of a breach of contract by either party, the IP or Huntex Ltd. This advice was given by KDLG, a major IP stakeholder, through the District Commercial Office (DCO).

The DCO initiated the process, had meetings with the IP members through their leaders to explain the implications of this registration, and went through the procedures with MTTI in Kampala. In 2013, the IP was finally registered as Bubaare IP Multipurpose Cooperative Society Ltd, the first among the 36 IAR4D-driven IPs in SSA to register as a cooperative society. The details of this registration as indicated on the certificate are shown in Appendix 4.1, while the Cooperative Society leaders are shown in Figure 4.4.

Why a Cooperative Society? In view of the fact that the IP was required to have legal status, the laws of Uganda provide four possible options of legal status: registration as a partnership, a company, a non-governmental organization

Figure 4.3 Bubaare IP Cooperative Society leaders: Julius Atuheire, IP chairman (squatting extreme right), Bertha Tushabe, IP Treasurer (standing extreme left) and Jeniffer Twebaze, IP Manager (standing in purple dress next to the pole)

Photo: CIAT/N. Russell

Figure 4.4 Dr Sospeter Nyamwaro, Project Coordinator, LKPLS, at the launch in
Bubaare

Photo: Bubaare IP

(NGO), and a cooperative society. The selected requirements for registration
under each of these are shown in Appendix 4.2. Given these provisions, the
IP could best be registered as a cooperative society. Bubaare IP Multipurpose
Cooperative Society finally launched its activities in June 2014 at a function
that brought together members of KDLG, partners, private sector and farmers
in Kabale District (see Figure 4.3).

Outcomes and impact of IP registration as a cooperative society

The registration of the IP as a cooperative society has created additional success
to the IP. Outcomes and impact can be observed in the following areas.

Infrastructural development

The sub-county authority had earlier recognized the developments at the IP,
appreciated the commitment of the members, and had donated a piece of land
at the sub-county headquarters for future developments. The sub-county had
also let the IP use a store that was not being used at Ihanga trading centre for
bulking and storing the sorghum. Registration of the IP cooperative society
has given the IP a new status. Considering the level of activity, the society has
decided to embark on the construction of a building to house their office, a

Figure 4.5 Bubaare IP Cooperative Society proposed construction plans
Photo: Bubaare IP

community bank, a potato processing unit, sorghum milling and packaging facility and a computer room. Figure 4.5 shows the construction plans for the building. The IP society will then have its own store at this building.

The premises at Huntex Ltd have also been expanded, and more equipment purchased to process larger quantities of sorghum (see Figure 4.6). The new equipment has the capacity to produce 2,000 litres of Mamera from 250 kg of sorghum per day compared to the previous capacity of 50 litres from 13 kg per day.

Figure 4.6 The expanded premises of Huntex Ltd (left) and new equipment purchased by Huntex Ltd (right)
Photos: Bubaare IP

Increased formation of Self Help Groups (SHGs) and membership in the society

Since registration of the IP cooperative society, there has been more SHGs formed and joining the IP to take advantage of the benefits provided by the society. Table 4.1 and Figure 4.7 show the number of groups and individuals that have joined the IP since it was established in 2009. Total membership has risen from a mere 32 in 2009, to 1,121 in 2014. The number of individuals however remained rather static between 2009 and 2011. This is because some farmers realized that the IP would not give them free inputs as they would have wished. Although a few farmers joined the IP, a few opted to drop out. The IP leaders took it upon themselves, during this time, to clarify to the community the operations of the IP and to emphasize that members would not receive free inputs. The farmers who appreciated this position joined the IP.

After becoming a cooperative, however, the new plans of the society provided incentives for more farmers to join. The plans included the possibility of signing a contract with Huntex Ltd to purchase more of the farmers' sorghum. This would ensure a market for their sorghum. The plans also included the possibility to internally mobilize funds to lend to member SHGs at a much lower interest rate than other microfinance institutions (MFIs). This sounded equally attractive so that after its registration, more farmers joined groups and the number of SHGs in the IP also increased.

Women farmers in particular have on average joined the IP in greater numbers than their male counterparts. This is most likely because the selected focus crop for the IP activities is a food crop, and commonly grown by women. After registration of the IP as a cooperative, the number of women farmers nearly doubled that of men farmers. Appendix 4.3 shows membership of men and women farmers in selected SHGs by the end of 2014. As a result of large numbers of women joining the IP, more of them have taken up leadership positions in their respective SHGs. In a sample of nine SHGs, three were found to have a woman as chairperson, five have a woman as vice-chairperson, five have a woman as secretary, four have a woman as treasurer and all have at least one woman as a committee member (Appendix 4.3).

Table 4.1 Membership, total number of groups and loan access between 2009 and 2014

	Year					
	2009	2010	2011	2012	2013	2014
Number of men	22	150	150	250	250	401
Number of women	10	250	250	250	500	720
Total number of farmers in IP groups	32	400	400	500	750	1,121
Total number of groups in IP	10	10	10	10	29	40
Groups accessing loans from IP	0	0	0	0	10	36

Source: IP file records, courtesy of David Tukahirwa, Secretary, Bubaare IP

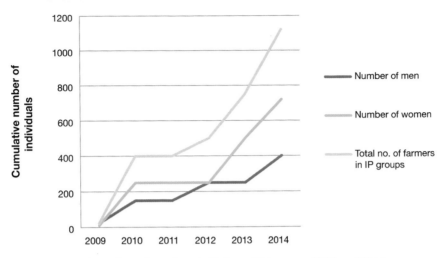

Figure 4.7 Total number of members in Bubaare IP between 2009 and 2014

Source: Bubaare IP

There has been increased access to small affordable loans

The society was able to internally generate funds from the members. Some 23 members committed themselves to put together UGX100,000 (USD36) each, to create a start-up capital for loaning to the groups. The funds were then loaned out to member SHGs at an interest rate of 1.5 per cent per month while other MFIs in the same area offer a rate of up to 4 per cent per month. The IP groups

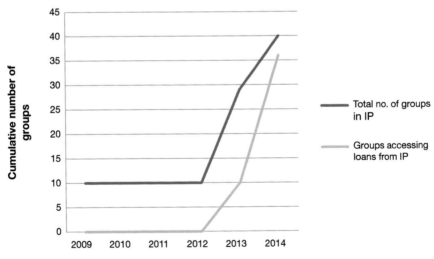

Figure 4.8 Total number of groups and loan access between 2009 and 2014 at Bubaare IP

Source: Bubaare IP

have since then borrowed between UGX600,000 and 3,200,000 (USD200 and 1,150), repayable in eight months. An individual may borrow as little as USD36 from the group to do a variety of activities of their choice. The funds are either used by the individual members to grow crops, or by an entire group to stock and store sorghum which is later sold when the market price is favourable. The funds can also be used for other personal enterprises. Examples of SHGs that received loans, and the use to which the funds were put, are shown in Appendix 4.4.

Table 4.1 and Figure 4.8 show the number of groups that received loans from the society. All ten groups that had joined the IP in 2009 received the first loans from the society in 2013. An additional 26 groups received loans in 2014. By the end of the year only four groups had not received a loan, which they did in January 2015. The IP cooperative society has therefore enabled access to small affordable loans that have opened up more market activities among the IP member groups and individuals.

There has been an improved supply of good sorghum grain to Huntex Ltd

The supply of sorghum grain to Huntex Ltd has since improved. Figure 4.9 shows the quantities of sorghum supplied by the IP for processing since 2011, increasing from less than 0.5 tonnes to 2 tonnes in 2014. For the period before 2011, small quantities were being supplied by individuals and were not recorded. However, after becoming a cooperative society, a contract was signed between the IP cooperative and Huntex Ltd for the society to supply specified

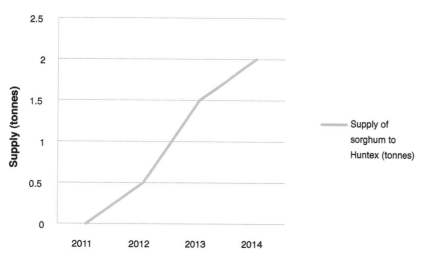

Figure 4.9 Supply of sorghum (tonnes) to Huntex Ltd by Bubaare IP Cooperative Society

Source: Bubaare IP

quantities at an agreed price. Both parties, the society and Huntex Ltd, have been able to meet their obligations.

> I now get the volumes I want, the variety I prefer, which is not adulterated. I purchase at a price that is more friendly to the IP than the open market and I am privileged to get supplies on credit.
>
> (Julius Byamukama, Manager, Huntex Ltd)

The demand by the SHGs for training has increased

The registration of the society is increasing the demand for training in savings and lending, marketing and processing, in anticipation of more market opportunities that the society might open up. KDLG in partnership with Makerere University were able to contract Durosh Empowerment Consult Ltd to train the SHGs in savings and lending (see Figure 4.10). A total of 32 groups were trained by the end of 2014 and were given a savings kit; a metallic box where they keep the groups' savings and records (see Figure 4.11). Women leaders in the groups have been particularly entrusted with the responsibility of counting the group savings every time the group meets to collect their savings.

The quality standards of the products are set to improve

The IP cooperative society has been able to link with the Uganda National Bureau of Standards (UNBS) through MTTI to obtain the S&Q marks for

Figure 4.10 IP members attending a training with Durosh in Kagarama Parish, Bubaare Sub-county

Photo: Bubaare IP

Figure 4.11 Group leaders receiving savings kits after training by Durosh Ltd in Nangara Parish, Bubaare Sub-county

Photo: Bubaare IP

Figure 4.12 Label showing the nutrient content of the sorghum flour

Photo: Bubaare IP

Figure 4.13 A 500g honey jar packed by Nyamweru Bee Keepers Association
Photo: Bafana Busani

quality certification. The Department of Food Science and Technology at Makerere University was able to teach the IP members to process high quality flour acceptable as food grade, and introduced the use of equipment such as food weighing machines, Sealers, food grade bags. The department has also analysed the nutrient content of the sorghum and produced a label for the sorghum flour packets (see Figure 4.12). This is one of the requirements for standard certification and it enables access to affluent markets in Kampala. Kampala city is eight hours away from Kabale and would otherwise be inaccessible to small farmers without this intervention.

The society is also pursuing the patenting of Mamera. Other value chains have similar developments; e.g. the National Organic Agriculture Movement in Uganda (NOGAMU) has picked interest in improving the quality of honey and promoting it in the regional markets (see Figure 4.13). The new status of the IP enables it to pursue these developments.

Underlying success factors

Short-term benefits of a small affordable loan

The IP cooperative has enabled farmers to obtain short-term benefits of small and affordable loans. The possibility of accessing such loans was a big incentive for farmers especially women to form SHGs and join the IP cooperative. Since they joined, the SHGs have all been able to get a loan although not all the members due to insufficient funds at the IP to cover all. However there is hope that they will all be covered as more funds become available.

BUBAARE IP MULTIPURPOSE COOP SOCIETY LIMITED (UGANDA)

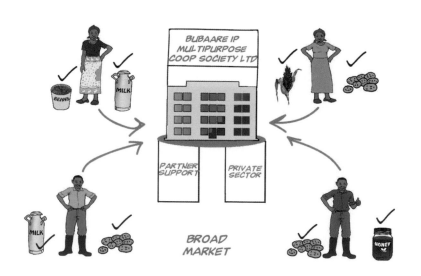

A wide distribution of the benefits

The small affordable loans are available to a wide range of socio-economic groups of farmers, who have the liberty to use the loans for a variety of enterprises. The farmers are also free to market their products outside the cooperative. Under normal circumstances, women and other poor farmers would not join a conventional cooperative society and would therefore not receive a loan from there. This is because such cooperatives would first of all not target their kind of enterprises. Second, women would not own items that they could present as security, and third, they would not have any formal identification. The IP multipurpose cooperative society has been able to overcome these limitations, enabling more socio-economic diversity in membership, and impact across the different socio-economic groups.

The new model of cooperative society

The IP multipurpose cooperative society is a new model of cooperative society. The new model is a key factor underlying success because it creates wider impact in the community with less transaction and monitoring costs than would have been the case with a conventional primary society. A conventional primary cooperative society is comprised of individual members who commonly produce one specific commodity such as coffee, milk or tea. These members have traditionally been small-scale farmers who were already producing a cash crop. The IP cooperative society on the other hand is comprised of SHGs as members. The groups are comprised of 20–30 individuals who under normal circumstances would not have been able to join a conventional society due to the absence of a relevant society for them in the area, or the lack of resources to buy shares and join one. The group members monitor each other's recovery of the acquired loan. This reduces the monitoring costs for the IP coop society. So far there have been no default cases as all loans due have been recovered in time.

The individual members of a SHG each produce commodities of their choice, at a scale that each can manage. The IP cooperative society therefore produces a variety of products from a very large number of small farmers, many of whom would not have been targeted by a conventional society. It is therefore a convenient model for large numbers of smallholder farmers. Furthermore, the IP society is only a part of the wider IP of which partners and private sector are also stakeholders, although not members of the cooperative society. This enables close association with, and continued support of partners such as research institutions, and the private sector together with the key stakeholders, the farmers. The IP remains open to this wide membership while owning its cooperative society. This is illustrated in Figure 4.14.

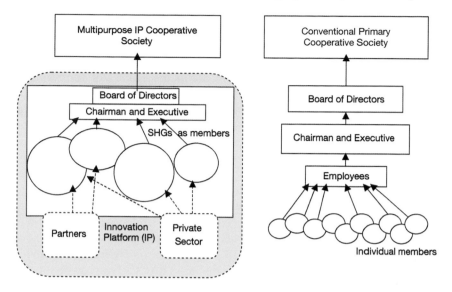

Figure 4.14 The structure of a multipurpose IP cooperative society versus a
conventional cooperative society

Source: Bubaare IP

Challenges that face the IP cooperative society

The major challenge at the moment is the need for the IP leaders and partners
to internalize the regulations of the new society, in order to assist the members
to operate within these regulations. The future of the IP cooperative society
seems promising although it gives rise to other challenges. It is not yet clear
how membership will be sustainably motivated. Members of conventional
cooperatives contribute shares, and at the end of the year receive bonuses
depending on the shares invested. The contribution of shares by member SHGs
has not yet been worked out in the IP society, as well as bonuses to be shared
at the end of the year. Further, as this society continues to expand, it might
have to hire professional employees to work under the executive (Figure 4.14)
so as to run the society as a business entity as successful conventional societies
do. This too, has not been worked out.

Implications for other IPs and development partners

The registration of Bubaare IP cooperative society has opened up opportunities
for a large number of smallholder farmers to participate in various market
activities. The IP that was established with a focus on the development of the
sorghum value chain realized increased production of sorghum grain, most of
which is purchased by the processor for making Mamera. The members are
able to process and pack sorghum flour, which has been introduced into nearby

supermarkets. Over the years, the IP has introduced other value chains; potatoes and honey. Besides, members enjoy the liberty of having other enterprises of their choice. The other three IPs in south-western Uganda that were established about the same time as the Bubaare IP have observed and appreciated these developments. The IPs are in the process of registering as cooperative societies.

Although Bubaare IP cooperative society is still young, we can draw some initial lessons for the three IPs and others in Uganda. First, an IP cooperative is a business entity with full legal rights. This is a status that gives it credibility, recognition, and more opportunities for support from development partners. This status will enable it to attract other credible partnerships for market opportunities both within and outside the country. Second, this arrangement leaves room for the IP to engage new development partners while still owning the cooperative. When it becomes necessary, the IP can engage the services of a consultancy company, a new processor, an input provider, among others, as new innovations develop. Third, an IP cooperative society itself can provide several services. The IP cooperative may operate a SACCO to provide suitable loans to its members, establish a bulking store for farmers' produce, operate a processing plant, and others as may be appropriate to address the complex challenges that rural smallholder farmers are always faced with. In view of these lessons learned from Bubaare IP cooperative society, other established IPs are encouraged to pursue this registration.

There are equally important lessons for development partners that associate with an IP cooperative. Private sector actors in particular need sufficient assurance that an undertaking will recover investment costs and remain profitable for some time. Under normal circumstances, this assurance cannot be guaranteed by small farmers, and so private sector actors do not make the kind of investments that the small farmers would need. In the Bubaare case, Huntex Ltd did not have the incentive and funds to expand their plant for processing sorghum, although locally grown. With the support of development partners on the IP, funds were made available for the expansion of the premises and purchase of equipment to process large quantities of sorghum under a contract. The willingness of the private sector to engage with the farmers can be enhanced if they are supported to make the investments required by the farmers in value chain development. Development partners with resources to make such investments should be encouraged to participate in the IP in order to support private sector engagement with smallholder farmers.

Lastly, it will be important for public sector institutions that are instrumental in facilitating the registration such as MTTI, and local governments, to be sensitized on the IP concept in order for them to help interpret the regulations of the cooperative movement for use in this model of cooperative society. Bubaare IP, being the first to register a multipurpose cooperative society, is still adapting itself to operating both as an IP and a cooperative society. On the other hand, the government institutions supporting the registration will have to work with more IPs requiring the same registration, yet with a diversity of activities and services. The sensitization of the public sector institutions on the IP concept will not only facilitate registration procedures but will enhance the transfer of the already observed impact of an IP cooperative society across the entire cooperative sub-sector in Uganda.

Conclusion

This case study demonstrates how the registration of Bubaare IP as a cooperative society has opened opportunities for a large number of smallholder farmers to participate in market activities. It has empowered the farmers into innovations and product diversification. Its mode of operation in particular favours women farmers who have responded in large numbers to take advantage of the benefits that it offers. Given a supportive environment, Bubaare IP promises to succeed as a cooperative society.

Acknowledgements

The following key informants who have contributed information for this case study are gratefully acknowledged by the authors:

- Mr Julius Atuheire, Current Chairman, Bubaare IP Cooperative Society.
- Mr David Tukahirwa, Secretary, Bubaare IP Cooperative Society, and Treasurer Nyamweru Bee-Keepers Association.
- Mrs Bertha Tushabe, Treasurer, Bubaare IP Cooperative Society, and Secretary, Ihanga-Hakona SHG.
- Mr Unity Rwanzigu, Secretary, Busirimuko SHG, Kagarama Parish.
- Mr James Tindikahwa, former Chairman Bubaare IP and retired Cooperative Assistant in charge of Rubanda County, Kabale District.
- Mr C.K. Gakibayo, District Commercial Officer (DCO), Kabale District.
- Mr Julius Byamukama, Manager, Huntex Ltd, Kabale District.
- Mr James Mugisha, Assistant Chief Adminstrative Officer (ACAO), in charge of Rubanda East County, KDLG.

The activities of the partners at the Bubaare IP were funded by the Sub-Saharan Africa Challenge Program of the Forum for Agricultural Research in Africa (FARA).

Appendices

Appendix 4.1 Details of registration

Name	Bubaare IP Multipurpose Cooperative Society Ltd
Registration number	10578/CRS
Location	Bubaare Sub-county Headquarters
District	Kabale District
Signed by	Registrar of Cooperative Societies, Kampala

Appendix 4.2 Legal provisions for registration of groups in Uganda

Legal form	Providing law	Selected requirements
Partnership	The Partnership Act, 2010	For business entities, limits the number of members to not more than 50
Company	The Companies Act, 2012	• A public company is government owned • A private company has membership limited to 100, tax obligations may not be suitable for farmer groups
NGO	The NGO Act, 2006	Non-profit making
Cooperative Society	The Cooperative Societies Act, Chap 112, 2012	Can register many members, recommended for farmers, fair tax obligations

Appendix 4.3 Selected Self Help Group (SHG) membership

	Name of group	No. of men	No. of women	Total membership	Size of committee	No. of women on committee	Positions of women★
1	Ihanga-Hakoona	11	12	23	7	4	1 sec, 3 cm
2	Busirimuko Kweterana	9	18	27	9	4	1 cp, 1 tr, 3 cm
3	Nyamweru Central	13	14	27	7	3	1 vc, 2 cm
4	Nyamweru Bee-Keepers Association	22	5	27	7	1	1 cm
5	Busirimuko Tukundane	14	14	28	9	5	1 vc, 1tr, 1sec, 2 cm
6	Busirimuko SHG	11	16	27	9	4	1 vc, 1tr, 2 cm
7	Karisabutungiburondwa	0	16	16	7	7	all women
8	Kitogota Farmers	0	23	23	7	7	all women
9	Kashenyi	9	17	26	9	2	1 tr, 1 sec
10	IP Cooperative Society	401	720	1,121	7	4	1 vc, 1 tr, 2 cm

★ Positions of women on the committee: cp–chairperson, vc–vice chairperson, sec–secretary, tr–treasurer, cm–committee member.

Source: SHG file records

Appendix 4.4 Selected SHGs that joined the IP cooperative society and received loans

1 Ihanga-Hakona SHG	2 Busirimuko SHG	3 Nyamweru Bee-Keepers Association
• formed in 2014; • has 23 members; • obtained UGX 600,000/= from the Cooperative Society in Dec 2014; • some six members have been loaned the money, UGX100,000 each; • some established nursery beds to sell vegetable and tree seedlings; • some purchased potatoes to re-sell, and others malted sorghum.	• formed in 2005, joined IP in 2009; • has 27 members; • was among the first 10 groups to obtain a loan from the IP in 2013; • the group has benefitted from the training in VS&L offered by Durosh Ltd; • loans used by individual members for various activities, such as expanding the production of sorghum.	• formed in 2014; • has 27 members; • obtained UGX1.4 m from the IP Cooperative Society in July 2014; • association has up to 1000 bee-hives; • members use funds for various enterprises including trade in potatoes and sorghum.
Bertha Tushabe, 50, Secretary, Ihanga-Hakona SHG	*Unity Rwanzigu, Secretary, Busirimuko SHG*	*David Tukahirwa, 58, Treasurer, Nyamweru Bee-keepers Association*

Note

1   USD1= UGX2,800.

References

Adekunle, A.A, Fatunbi, A.O., Buruchara, R., Nyamwaro, S., 2013. Integrated agricultural research for development: From concept to practice. Forum for Agricultural Research in Africa (FARA), Accra, Ghana.

Kwapong, N.A., 2013. Restructured agricultural cooperative marketing system in Uganda: Study of the 'Tripartite Cooperative Model'. Euricse Working Paper no. 57, 13.

Kwapong, N.A., Korugendo, P.L., 2010. Revival of agricultural cooperatives in Uganda. Uganda Strategic Support Program (USSP), Policy Note no.10. International Food Policy Research Institute (IFPRI), Washington D.C.

Selected Acts from the Laws of Uganda: The Companies Act, 2012, The Cooperative Societies Act, Chap. 112, 2012, The Partnership Act, 2010, and the NGO Act, 2006.

The Cooperative Society Regulations (CSR), 1993.

5 Crop–livestock–tree integration in Uganda

The case of Mukono–Wakiso innovation platform

Sylvia Namazzi, Perez Muchunguzi, Dieuwke Lamers, Anna Sole-Amat, Piet van Asten, Thomas Dubois, Victor Afari-Sefa, Moses M. Tenywa, Immaculate Mugisa, Mariëtte McCampbell and Murat Sartas

Introduction

In the Lake Victoria Crescent Zone of central Uganda, farmers have over recent years been struggling with increasing poverty and malnutrition, due largely to low agricultural yields. The farmers wake up every day to a battle of pests, diseases, fake agricultural inputs, poor access to markets, post-harvest losses and infertile soils. Farm production in this region is rain-fed and is already being hurt by climatic changes. In 2013, the CGIAR research programme Humidtropics initiated innovation platforms (IPs) in Uganda to join the farmers in their fight. Four IPs (Mukono–Wakiso, Kiboga–Kyankwanzi, Luwero–Nakaseke and Masaka–Rakai) were presented. However, only Mukono–Wakiso and Kiboga–Kyankwanzi started operating. The Mukono–Wakiso platform focused on farmers' production and marketing while simultaneously strengthening their local capacities. Many organizations were involved in this work and they faced many pitfalls. Platform actors were dropping off along the way, some of the international research organizations involved gave the platform too little attention, and organizations tended to invest only in their preferred crops and commodities rather than the integration of several commodities, as the farmers wanted. This case study explores how the IP is tackling these and other problems to make a difference in the lives of the Mukono–Wakiso farmers.

The study site

Mukono and Wakiso are two districts where Humidtropics has one of its field sites in the Lake Victoria Crescent Zone of Uganda. This peri-urban area is highly populated, reducing the acreage under agriculture and threatening food security. The two districts have a booming business of real estate; many city entrants are opting to construct houses in either Mukono or Wakiso district. This increases competition for the land uses and exerts pressure on natural resources such as forests and wetlands. The districts are close to Kampala, the capital of Uganda, providing a big market opportunity for the farmers' produce. Despite the business opportunities, the two districts have high youth unemployment close to the national average of 62 per cent, posing a challenge and opportunity for the IP (UBOS, 2002).

The birth of Mukono–Wakiso IP

From the national Humidtropics inception meeting in August 2013, a local non-governmental organization – Volunteer Efforts for Development Concern (VEDCO) – volunteered to lead the other partners in forming the platform. In early 2014, VEDCO, International Institute for Tropical Agriculture (IITA), National Agriculture Research Organization (NARO) through Mukono Zonal Agricultural Research Institute (MUZARDI), and Makerere University started the initiatives to form the Mukono–Wakiso platform along with the local government.

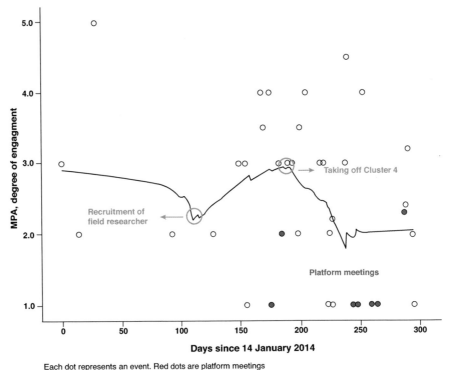

Each dot represents an event. Red dots are platform meetings

Figure 5.1 Line graph showing the dynamics in participation on the platform

Source: Sartas *et al.* (2015)

All the active actors in Mukono and Wakiso districts were invited for an inception meeting that resulted in the birth of the Mukono–Wakiso platform. The participants at the meeting together decided what positions were required on the platform and who could best fit these positions. The platform stakeholders who included civil society organizations, research organizations, private sector, local government, academia and extension chose an executive team. The executive team is led by a chairman from Mukono local government office while the deputy chairlady is from Wakiso local government office.

From then forth the platform engaged in various activities in a dynamic way (see Figure 5.1). The platform is open to any stakeholder that is interested. Consequently, new members came on board (e.g. AVRDC-World Vegetable Center, Kampala Capital City Authority and Farmgain Africa), while others dropped out. This was particularly the case for some private sector actors when they did not see an immediate 'business case' for their product. Interesting to note is that the new members often come as a result of good ambassadorship.

The first meetings of the platform were used to agree on common goals, division of tasks, and ways forward, including the identification of entry points that would respond to farmer needs. To date, the platform holds meetings on a monthly basis to share lessons and figure out how to tackle farmers' challenges. Additional meetings are sometimes convened by the chairman on a need-be basis, leading to two or more meetings in a week. The chairman in addition organizes study tours/exposure visits for the farmers, and attracts additional (government) funding into the platform.

Where does the platform begin?

Selection and validation of entry points

A facilitator from Makerere University guided the platform members to select pressing needs of the farmers in line with the Humidtropics intermediary development outcomes (IDOs). The members were grouped under each IDO to identify the three most pressing farmer needs. The platform members engaged in group discussions to come up with issues that pointed to (i) limited land, (ii) declining soil fertility, and (iii) climate change among other problems (see Box 5.1).

After identification of pressing farmer issues, the platform members asked: 'What system combinations should we take on to address the farmers' issues?' The IP identified up to nine different systems:

1   Banana + Coffee + Vegetable + Agro forestry
2   Banana + Coffee + Vegetable + Agro forestry + Dairy
3   Banana + Sweet Potato + Piggery + Agro forestry
4   Vegetable + Poultry + Piggery
5   Banana + Coffee + Vegetables + Dairy + Maize
6   Vegetable + Poultry
7   Banana + Vegetables + Poultry + Agro forestry (fruit trees)
8   Poultry + Maize + Vegetables
9   Banana + Dairy + Poultry + Vegetables + Beans.

Box 5.1 An excerpt from platform formation report showing identified needs of farmers from Mukono and Wakiso

Group I: FOOD SECURITY

The three most pressing issues

- Reducing land for agriculture
- Minimum participation of the productive forces
- Unpredictable weather patterns

Group II: NUTRITION AND HEALTH

The three most pressing issues

- Feeding patterns/habits, preparation methods, knowledge gap
- Climate change
- Urban market drive to focus on high value enterprise

Group III: POVERTY REDUCTION

The three most pressing issues

- Use of traditional methods of production (subsistence agriculture versus commercial)

Group on food security identifying the pressing issues
Photo: Mukono–Wakiso IP

Box 5.1 *continued*

- Lack of collective marketing systems by farmers + Poor infrastructure
- Lack of clear policies urban agriculture

Group IV: NATURAL RESOURCE MANAGEMENT

The three most pressing issues

- Loss of soil fertility
- Poor enforcement of regulations
- Climate change

Source: Mukono–Wakiso platform minutes

The platform analysed each of the systems and agreed on an integrated system of crops, livestock and trees that includes banana, vegetable, poultry, agro forestry (emphasizing fruit trees) and piggery. Vegetables featured strongly in almost all identified systems, for the primary reason that they mature and sell quickly, making them a fast income-generating activity. Additionally, vegetables were also identified to provide opportunities for women and youth in the districts. Thereafter, CGIAR centres together with the local organizations on the platform carried out a rapid survey among the farmers to validate the integrated system selected.

Why system integration?

An integrated farming system is an often diversified agricultural production system that seeks to effectively link all farm enterprises to improve the efficiency of land, labour, finance and nutrient investments. It consists of a range of resource-saving practices that aim to achieve acceptable profits and sustain production levels, while minimizing the negative effects of intensive farming and preserving the environment (Rota and Sperandini, 2010). Simultaneously, it provides opportunities to strengthen the resilience and sustainability of farmer livelihoods.

The Lake Victoria Crescent Zone in which Mukono and Wakiso are located is largely dependent on a banana–coffee farming system (Sserunkuuma, 2001; Van Asten *et al.*, 2011). Yet productivity and the relative importance of this system has significantly reduced over the past decades. Development projects and extension officers have largely adopted a value chain approach, thereby encouraging farmers to specialize in sole crop production in a bid to increase quantities for sale. Unfortunately, climate change is increasing the animal and crop disease/pest burden in the area, rendering farmers' enterprises vulnerable to climate and price shocks.

In an integrated system livestock, crops and trees are produced within a coordinated framework (Figure 5.2). The waste products of one component serve as a resource for the other. Therefore, the incorporation of livestock, crops and tree farming systems provides an opportunity to improve sustainable access to income and nutrition by spreading risks. This also sustains the natural resource base through nutrient recycling, erosion control and pollination services, among others. Due to the limited farm size in Mukono and Wakiso, livestock is often kept under zero-grazing and farmers rely on additional feeds/fodder from outside the farm, at least for part of the year. This in the end makes livestock-keeping an expensive enterprise. However, improving on farm fodder availability throughout the year is feasible, e.g. by planting fodder trees such as *Calliandra* and through re-use of crop residues from vegetables, banana and sweet potato among others. In this regard the platform is involving different partners to respond to the knowledge requirements of farmers on how best to use their existing on-farm resources to feed their animals and cater for their energy needs.

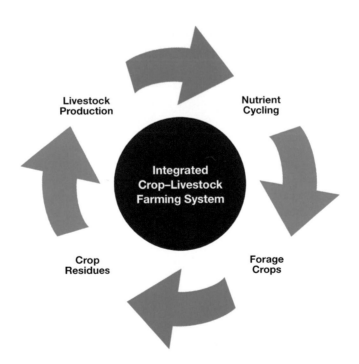

Figure 5.2 Integrated crop–livestock farming system – key aspects

Source: Rota and Sperandini (2010)

Managing different systems on a small plot is a challenge for the farmer. Practices such as rotation, soil fertility management, water conservation, waste management and animal nutrition compete for the limited land, labour and capital that farmers have available. Hence, researching what system works best for farmers is an opportunity for the platform to engage in and requires the involvement of different research organizations to answer farmers' questions through research that is relevant to the platform.

Platform efforts to answer farmers' questions

Going to the field!

A series of research for development (R4D) activities were initiated in response to the platform needs (see Figure 5.3). These included (i) situation analysis, (ii) baseline survey, (iii) market survey and (iv) agro-biodiversity survey. These activities were done by platform members together with the CGIARs to obtain baseline information before implementation of major activities to properly monitor changes among farmers.

The platform started field activities with trees provided by ICRAF working together with a local organization (Ssaza Kyagwe). The agroforestry trees selected were *Albizia*, *Graveria*, *Eucalyptus* and mangoes. These trees were selected for their potential to contribute to soil fertility improvement, provision of fodder, timber and fruits. *Eucalyptus* was preferred because it provided fast income for the farmers.

Figure 5.3 Farmers going to one of the field trials during a training
Photo: Mukono–Wakiso IP

Figure 5.4 Farmers in the field identifying pests and diseases
Photo: Team Uganda

A few farmers from Mukono were involved in this activity. This was because at the beginning, the platform was leveraging on ongoing CGIAR activities in the field sites. In addition, farmers from both Mukono and Wakiso were provided access to (indigenous) vegetables from AVRDC with the help of the local government (see Figure 5.4).

Research experiments

The platform conducted research on integrated soil fertility management (ISFM) to understand how farmers can utilize manure and crop residues to improve soil productivity. Bio-slurry and chicken manure are evaluated to validate recommended rates as well as determine the response of various vegetables to each manure (see Figure 5.5). Uganda Christian University (UCU), a member on the platform, is leading other partners in finding solutions using on station experiments for the platform to better advise the farmers.

The experimental research and some other activities were funded with resources from the Humidtropics platform-led innovation funds (also referred to as Cluster 4).

Farmers go to class! Training of farmers

The platform creatively devised means to build the capacity of the farmers in their respective groups to effectively engage in different activities (see Figure 5.6). The platform organized trainings on integrated systems operation at production level to empower the farmers to identify synergies among the different activities they engage in at plot and farm levels.

Figure 5.5 UCU ISFM experiment on Nakati

Photo: Team Uganda

Figure 5.6 Farmers attending a training on systems integration

Photo: Team Uganda

The farmers have been trained in value addition and marketing to help them manage their businesses better. They have also been trained in business planning to help them get ideas of how they can move their enterprises into money making ventures and increase profitability (see Box 5.2). The vice-chair lady of the platform lobbied successfully for an irrigation training facilitated by an organization (Vibes) that was not part of the platform. This training equipped farmers with skills to continue production during the dry spell particularly, using water from fish ponds.

This training was particularly important for the peri-urban vegetable farmers whose vegetables fetch a better market price in the dry season. In addition, farmers were trained on nutrition to emphasize the importance of eating a balanced diet and to encourage them to integrate the crop–livestock–tree system to increase access to the different foods required.

Linking farmers to markets

Markets are still a challenge for the farmers even when they are close to the capital city. The middle men offer very low prices to the farmers and yet they do not allow them to penetrate the markets once they make efforts to. Markets are disorganized and there are weak linkages between the producers and the buyers. Perishability of most of the agricultural products forces farmers to sell cheaply to middle men for fear of produce rot and wastage. The platform tasked Farmgain Africa (a member of the platform) to link farmers to markets.

Farmgain Africa is a consultancy firm that specializes in agri-business, market information and agro enterprise development. The organization is identifying traders that can work with the farmers at the available scale and capacity to provide a market for their different products. The organization collaborates with the farmers and has encouraged them to utilize the market opportunities around them before venturing outside. This helped to open farmers' eyes and currently the farmers from Wakiso are buying sweet potato vines for planting from Mukono farmers. One farmer was heard saying: 'We bought sweet potato vines from Mukono farmers. The last time we met in a training, we exchanged contacts with them. Recently we bought three bags of vines from them.'

Farmgain Africa has engaged with the farmers by visiting their groups to understand their market arrangements/models in order to identify gaps. The organization found out that piggery and poultry have a different market arrangement from vegetables. The vegetables are sold in groups with middle-men and transporters involved, while for piggery and poultry products the buyers know whom and where to buy from. The organization is finding out how it can draw up memoranda of understanding or contractual arrangements between farmers and big buyers on behalf of the platform.

Meetings with partners to brainstorm on markets are being held to understand short-term and long-term market strategies for the farmers' products. Traders (middle men) have been engaged in platform market activities to see

IMPROVE QUALITY PRODUCTIVITY
MUKONO-WAKISO IP (UGANDA)

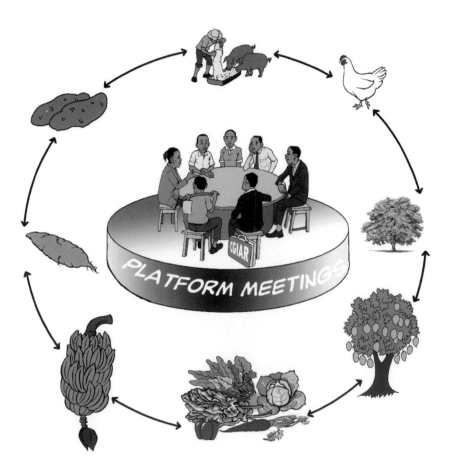

Box 5.2 Case of two farmers in Mukono integrating crop, livestock and trees

Mr Kigoonya Augustine is a young farmer married to Susan Kigoonya with two baby boys. They live in Naggalama, Mukono district. He is the chairman of Tukolerewamu farmers' group which is made up of 30 farmers engaged in production of different crops and animals.

He was working as a shop attendant in Ntinda, a suburb of Kampala, where he was earning 300,000 UGX per month. He said, 'I was spending all the money on food, transport and rent and would have to run back to my father for some financial help'. He could not cope with this any longer and decided to leave Kampala for Mukono to do farming. He started with sole crops of beans which he lost due to seasonal changes, then tried maize sole crops from which he did not reap much due to price fluctuation; he later decided to diversify with a number of crops, i.e. bananas, tomato, maize, sweet potato and beans. With the diversification he is reaping more.

Mr Kigoonya said, 'I now earn 1.5–2M UGX as income at the end of one season. I do not spend on food because I have banana, maize, sweet potato and vegetables to sustain my family. I get manure from my father's

Mr Kigoonya harvesting his vegetables
Photo: Team Uganda

Box 5.2 *continued*

poultry house but am hoping to own chickens to have ready access to manure. We will not be buying eggs and my money will go up. I have also obtained knowledge of system integration and natural resource management from Humidtropics trainings and I have decided to get my own poultry to better use my fodder from the field and have eggs and manure there.' He says the only vegetable they participated in before was tomato but with the Humidtropics they are able to grow various vegetables for home consumption and sale.

He has incorporated fodder trees such as *Calliandra* in his plot to help with nutrient recycling and is currently using the fodder for mulching the vegetables and bananas.

Ms Beckie Nakabugo is a professional designer trained in Nairobi, Kenya but loves farming to the annoyance of her father who wanted to see her prosper in white collar jobs rather than farming. Beckie (in the same group as Mr Kigoonya) worked in a forex bureau before but left and started farming, using her savings. She started well and a piggery was her first choice. Everything was rosy and everyone looked up to her as she raised

Beckie engaging the IITA Board of Trustees in the field
Photo: Team Uganda

Box 5.2 *continued*

pigs, up to 100 of them! In a peri-urban area, this is a very big achievement. Swine fever came and swept through all of them. She was devastated but wanted to continue farming so she decided to diversify with different enterprises in order to spread her risks.

She said, 'at the beginning I got differing advice from people on what enterprise to start with and that is how I ended up starting a piggery. But all this information was just pointing at how much money I get at the end of the day without telling me about the challenges.' Beckie continued: 'by the way, I also tried ground nuts because someone told me that a kilogram is 4,000 UGX so in my calculation, I realized that I could get a lot of money if I got only 100 kgs. But when it came to shelling the ground nuts, it was disaster! The people who are hired to shell, each goes away with half a kilo in their clothes and if it is ten people for a few days that's so many kilograms lost. But when I had Humidtropics training, especially the one on business planning, I learned a lot of things.'

'First of all, I have learned to love farming, not looking at it as a punishment. Now I know how to "*kubalirila*"', she laughs as she translates it into English, '*calculate profitability* of my business. And also to invest and get money from my enterprise. I used to enjoy looking at my pigs increasing in number; just seeing how big and many they were gave me a smile and I did not want to sell them and yet I would have sold them and invested again. I have also liked the interactions with the other farmers under Humidtropics. One farmer dealing in tomato told me that a certain variety (*Asila*) does not perform well on newly opened ground. Had I met these farmers before, I would not have made the loss that I incurred after buying 4 kgs of Asila each at 500,000 UGX and none of these germinated partly because I mixed the fertilizer with the seed at planting. Actually, Humidtropics has helped us get knowledge; many of us do not know the technicalities of farming and yet the agro-shop attendants too are ignorant about the specifications of the inputs they sell to us.' Beckie was still narrating more of her experiences since she had tried various enterprises and had met a number of challenges but the interview had to come to an end.

how the middle men can offer farmers better prices for their produce. The discussions held are captured in minutes, which are then shared via Dropbox.

Moreover, strategies to link the Mukono–Wakiso platform and the Kiboga–Kyankwanzi platform are underway to allow exchange of produce from the different Humidtropics platforms. For example Mukono–Wakiso farmers could buy soybean and maize from Kiboga–Kyankwanzi farmers for use as animal feed.

Conclusions

Positive results from Mukono–Wakiso platform

In the past year, the Mukono–Wakiso platform worked through different institutions to answer farmers' needs. Through its engagements the platform identified vegetables as the starting point and this resulted in the lobbying for the coming of the AVRDC–to Uganda. When the platform selected vegetables as a central part, it realized that none of the CGIARs centres on the platform had expertise in this area and that is how the lobbying for AVRDC was started.

In the meantime, the vegetable activities on the platform have increased the participation of youth in platform activities. This has demonstrated that providing specific inputs pushes activities for quick wins for the farmers and platform members. Young people are going back to farming in central Uganda because of the quick money from vegetables and the platform has been instrumental in encouraging them. The IITA board of Trustees (BoT) visited the platform and selected two youths to attend the 'agripreneur' youth training in Ibadan, Nigeria (Figure 5.7). This is leading to the creation of a fully fledged youth programme in Uganda.

The platform efforts to help farmers link to markets has increased the interest of various stakeholders in platform activities. For example, Farmgain Africa, a private organization, has picked keen interest in the platform. A member from Farmgain noted that the platform has made them take on linking farmers to markets as one of their areas of operation which was not the case before joining the platform. They are considering working closely with farmers besides those under Humidtropics and connecting them to markets. Other organizations are looking at the opportunities that are arising from the market-led processes and are increasingly engaging in the platform with full energy. Likewise, the farmers are hopeful because they have always had the problem of markets but the platform is helping them find solutions. As a result, farmers are participating fully in all platform activities at all times. Also, the operations of the platform in the two districts of Mukono and Wakiso improved networking for the farmers and increased market opportunities for them. Moreover, communication between platform stakeholders is improving. For example, the platform members have a WhatsApp group on which they share information and events. This is all happening amidst various challenges.

Challenges encountered by the platform

The success of this platform is limited by various constraints. Harmonizing the different actors to participate actively and continuously in the integration process is still daunting. It is a challenge for the platform to create win–win situations for more private sector engagement. Therefore, private sector involvement on the platform is still low. Private organizations are not interested in sitting in

Figure 5.7 IITA BoT team in Mr Kigoonya's field (top) and youths attending one of
the IP trainings (bottom)

Photos: Team Uganda

regular meetings but rather prefer to participate on a particular issue and leave.
Likewise, each international research organization has its own core research
areas and still wants to contribute to that. It thus becomes difficult to have them
pick interest in the platform activities let alone support the activities with
funding. The local partner organizations who come to meetings often expect
a transport refund that when not provided reduces their motivation to
participate.

The financial burden of facilitating the platform increases with such demands
as transport refunds. Whenever meetings are convened the platform covers costs

related to teas and lunches at the meetings. This was going well in the beginning when the facilitation came from the Humidtropics office but once this is delayed or removed, the platform does not have a clear source of funding for this purpose.

In addition, focusing on all the different commodities at the same time is still a challenge. For example, when discussing vegetables one variety (*Nakati*) dominates. The integration with poultry and piggery still lies in wait on the platform. Funding too, for the different integrations is a challenge. Farmers without livestock are interested in livestock incorporation in their cropping system but no organization is out to provide these, as was the case with AVRDC and ICRAF who provided seed kits and tree seedlings respectively.

Any lessons from Mukono–Wakiso IP?

Lessons from this IP are dynamic. Focusing on one major commodity that helps pull other commodities is an important lesson from the platform. The vegetables took a central place on this platform; but as the IP dealt with the vegetables, poultry and other animal-keeping activities became attractive because of the vegetables' heavy need for manure. Yet manure from other sources is expensive and limited in availability. Fodder trees, then, become linked to feed the animals and as well provide mulch for the vegetables.

Reflection meetings by a small team (referred to as secretariat) to guide the activities of the platform have been crucial for this IP. These meetings are held monthly by a team of facilitators to assess the platform process. It is through these meetings that work plans are drawn and lessons from previous months used to guide the current month. Individuals are tasked to follow up on different activities being run by the participating organizations. These meetings give the IP better guidance to move forward.

The platform has learned the importance of bringing actors together only when needed to avoid fatiguing them with meetings without tangible benefits. Partners in the private sector are slowly being brought back to the platform in this way. Also, specialized meetings are held with a concerned actor to get feedback that nourishes the platform. This is usually done by the secretariat and the platform chairman.

Not much would have been arrived at by this platform if it were not for the direction and facilitation from the leaders. The chairman has been instrumental in guiding this platform in the past year. He at one point suspended regular meetings until there was activity engaging farmers on the ground. The platform facilitator (Action Site Facilitator) too, has encouraged innovative ideas to emerge from all participating members. He inspires free discussions on the platform and because of that he prefers people to call him by his first name instead of professor (the facilitator is a professor of soil science from Makerere University).

It has also been noted on this platform that processes driven by markets are useful in fostering partnerships and networking. It is from this that some private sector actors are finding the platform useful. The farmers are also greatly involved because of the promise they foresee.

Assigning tasks to partners on the platform makes them feel relevant and committed to the platform. Partners such as Farmgain, VEDCO and UCU were assigned roles on the platform. In that way their participation on the platform increased and as a result, other partners also want to lead in some activities. This was the case when Cluster 4 funds were distributed to partners to lead various platform activities.

What next for Mukono–Wakiso?

The integrations being implemented by some farmers still need a lot of fine-tuning, highlighting the need for good systems demand-driven research to continuously give clear recommendations. The interest in the integration of the different scaling partners on the platforms is also growing, thus presenting a good opportunity for the farmers, while creating awareness too.

To be able to have the desirable impact, the platform has decided to have a high level delegation (R4D platform at national level) that includes ministers as part of the scaling team. This is to increase policy engagement and fundraising. It is envisaged that engaging policy makers at national level will drive the platform better.

While most funding opportunities usually target value chain-based projects, partner organizations within the platforms have vowed to come together to develop joint proposals for funding in order to expand the systems integration processes' research.

The platform is working towards strengthening of monitoring and evaluation to enable multilevel data capturing to track the changes in knowledge, attitudes and skills throughout the platform process. Several activities are happening on the platform and there is no system in place to capture all that is going on. The platform therefore is embarking on tracking all the activities and events that are happening, to fast track its performance and report better to the rest of the world.

Acknowledgements

This platform was formed and facilitated under the CGIAR Research Program Humidtropics. Humidtropics also provided funding for this case study. The chairman of the platform and other members of the platform gave instructive contributions towards the development of this case study.

References

Rota, A., Sperandini, S., 2010. Integrated crop–livestock farming systems. Livestock Thematic Papers. International Fund for Agricultural Development (IFAD), Rome, Italy. www.ifad.org/lrkm/factsheet/integratedcrop.pdf. Accessed 15 September 2015.

Sartas, M., Muchunguzi, P., Sole, A., Lamers, D., Namazzi, S., Awori, M., Schut, M., Tenywa, M., Leeuwis, C., van Asten, P., 2015. The stepping stones to success: How we achieve high ownership and reflective learning in multi-stakeholder processes in Uganda? A poster prepared for the International Conference on Integrated Systems Research, March 2015, Ibadan, Nigeria. www.researchgate.net/publication/2727 70276_The_Stepping_Stones_to_Success_How_We_Achieve_High_Ownership_and_ Reflective_Learning_in_Multistakeholder_Processes_in_Uganda. Accessed 27 February 2015.

Sserunkuuma, D., 2001. Land management problems and potentials in the lakeshore intensive banana–coffee farming system. A paper prepared for the Regional Conference on Policies for Sustainable Land Management in the East African Highlands held 24–26 April 2000 at the United Nations Economic Commission for Africa Conference Centre, Addis Ababa, Ethiopia. http://ussp.ifpri.info/files/2011/10/land-management-in-banana-coffee-system-sserunkuuma-2001.pdf. Accessed 12 March 2015.

UBOS, 2002. *National population and housing census.* Uganda Bureau of Statistics (UBOS), Kampala, Uganda.

Van Asten, P., Wairegi, L.W.I., Mukasa, D., Uringi, N.O., 2011. Agronomic and economic benefits of coffee–banana intercropping in Uganda's smallholder farming systems. *Agricultural Systems* 104, 326–334.

6 Humidtropics innovation platform case study

WeRATE operations in West Kenya

*Paul L. Woomer, Welissa Mulei and
Celister Kaleha*

Origins of WeRATE

WeRATE is built upon a common understanding that isolated farmer groups and local NGOs cannot satisfy the expectations of their clients unless they work together to exchange ideas and opportunities. This realization was slow to emerge as local organizations were often territorial and secretive, believing it was in their best interest to seek and work with sponsors independently. It was the emergence of umbrella organizations, such as WeRATE that demonstrated the advantages of collective action to these smaller local groups. For example, WeRATE members interviewed in February 2015 declared: 'WeRATE has helped in facilitation, training and dissemination of how to use technologies and value addition'; 'Value addition such as processing has empowered women to earn money, WeRATE has also opened up markets in Nairobi and villages' (Appendix 6.2).

Innovation

WeRATE has collaborated in the development of several innovations, both among its members and in collaboration with the private sector. WeRATE demonstrated the efficacy of IR maize as an effective tool to combat striga and incorporate this technology into an integrated control system. Following these guidelines, WeRATE farmers were the first in Africa to eliminate striga from their fields and farms. WeRATE demonstrated the advantages of marketing BIOFIX legume inoculants in packets smaller than 100 g so that this product better reflected the demands of small-scale farmers. Now inoculants are also available in 10, 20 and 50 g packets. WeRATE pioneered soybean enterprise throughout West Kenya, first introducing more productive varieties, assembling BNF technologies, introducing them to agro-dealers and then overcoming the emergence of Asian rust disease through the introduction of tolerant varieties. WeRATE worked with MEA Ltd to formulate a new, widely popular fertilizer-blend (Sympal), specially blended for symbiotic grain legumes. WeRATE led in the development of recipes using soybean so that the nutritional advantages of this new crop would not bypass the households otherwise adopting soybean grown for market. WeRATE is sensitizing farming communities to the threat of invading Maize Lethal Necrosis Virus and working with farmers to develop non-host alternatives and promote tolerant maize varieties. To identify a single major innovative accomplishment by WeRATE is superficial as the true strength of the IP is its ability to work with both researchers and farmers in a practical, iterative problem-solving mode.

From umbrella organization to IP

WeRATE operated as an informal network in West Kenya for many years before it was formalized as a registered NGO. First it operated through consensus among NGOs active in Kenya's Western Province dating from the mid-1990s through 2002 with modifications to the initial approach as more NGOs joined the network (Woomer *et al.*, 2002; Woomer, 2007). Moi University began research on small-scale farming systems and joined this informal alliance (Okalebo *et al.*, 2006). At this point, about 240 on-farm technology trials were being conducted per year. FORMAT was formed in 2002 by MSc graduates from these projects and the term WeRATE was first coined as the outreach arm of that NGO (Savala *et al.*, 2003). Its approach in collaboration with AATF and later Alliance for a Green Revolution in Africa (AGRA) was then applied to other farm technologies, particularly the management of striga (AATF, 2006). In 2008, WeRATE reached about 52,000

Figure 6.1 Participants at the WeRATE Planning Workshop for the 2015 long rains and Second Agricultural Technology Clearinghouse

Photo: WeRATE

households for pre-release testing of imazapyr-resistant (IR) maize for control of striga (Woomer *et al.*, 2008), a 'miracle' technology later commercialized by three Kenya seed companies. In 2010, the N2Africa Project identified WeRATE as the lead outreach partner in the Western Kenya Action Site (WKAS), and a system for geographic 'Nodes' coordinating 26 cooperators was developed; node leaders were responsible for administration and logistics as several new, large farmer associations had emerged. It forged close working relations with several companies that manufacture and distribute farm inputs, particularly MEA Fertilizers Ltd, SeedCo Kenya and the Kenya Agro-dealer Association. A photo of WeRATE members appears in Figure 6.1.

Formalization and operations

In 2012, the Humidtropics programme sought collaboration with Research for Development (R4D) Platforms for intervention and possible resource transfers in its WKAS (Figure 6.2). WeRATE's bid for this position was successful. At the same time, Phase 2 of the N2Africa Project entered into 'indirect' technology outreach, meaning that field actions could no longer be coordinated directly by IITA.

In addition, several new initiatives were seeking outreach partnership in areas of soil fertility management, bean disease control and improved cassava-based cropping. Formalization of WeRATE as an umbrella NGO was initiated at the final N2Africa Kenya Country Workshop in February 2013 and the NGO was officially recognized, starting 23 May 2014. WeRATE's main objective is to advance rural transformation in West Kenya. It was formed in part to become eligible to receive funds directly from donors and become equal partners in larger scale research and development activities. After a lengthy approval process by the NGO Board of Kenya, WeRATE now has its own bank accounts (both US$ and KES) and a KRA Pin Number. As a result, it will no longer manage funds through member accounts. This should lead to better financial reporting to supporters. Its officers include: Chairman, Vice Chairman, Secretary, Treasurer, M&E Specialist, Extension Specialist, Data Manager, Accountant and four Technical Advisors. Only registered organizations with an email contact and paid membership dues of KSh3,000 (about $35) were eligible for participation; 22 groups, mostly farmer associations and local NGOs, met these criteria (two more joined later). These groups, their areas of operation, farmer representation and activities are further described in Appendix 6.1. A map of the WKAS and location of WeRATE members within it appears in Appendix 6.3.

Farming systems

Farming systems within WKAS were characterized through a comprehensive survey of 291 households conducted by the N2Africa Project in 2013. Overall farming system and household characteristics were integrated into a farming

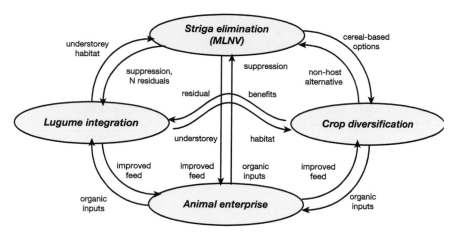

Figure 6.2 Key entry points for intervention and their possible resource transfers in small scale farming systems adopted by WeRATE and the Humidtropics programme in the West Kenya Action Site

Source: WeRATE research

system diagram that depicts crop and animal enterprises, resource transfers, household food supply and income (Figure 6.2). Average farm size is rather small (0.87 ha), farm activities with maize and bean intercropping predominate, less importance is placed upon root and cash crops, farm management practices indicate that crop residues are frequently being transferred between fields, fed to livestock and used in composting. Least common practices include top-dressing with mineral N, mulching or transfer of fresh manure and urine, all managements with proven efficacy (Sanginga and Woomer, 2009). This approach suggests that income from sales of cereals, legumes and animals constitutes 76 per cent of the household income per year. However, some elements of this model are based upon outside information (e.g. commodity prices) and assumptions (crop residues = 1 − Harvest Index) and some resource flows are absent for lack of information. The findings and analyses of these farming systems provide a strong baseline and perspective upon which to base future innovative and R4D actions.

WeRATE also conducted a survey among its member groups in late 2014 to determine their activities, capacities and needs. A 24-query questionnaire was developed and administered to 25 stakeholder groups. Results showed that altogether these stakeholder groups represent 79,506 farmers, 66 per cent of whom are women, there is strong interest among these groups to better understand and access new farm technologies, youth and women interests are strongly represented. During 2014, 86 field trials and 36 farmer field days were conducted by WeRATE members. Farmer grass-roots training is also a priority among these groups with 6,265 members (58 per cent women) trained in various technologies; WeRATE popularized itself and promising technologies through media events in 2014. A majority of members operate their own input shops but also work closely with other agro-dealers, produce seed and conduct collective marketing, with 75 tons and 182 tons produced and distributed respectively, directly engaging 7,645 members (70 per cent women). Value added processing is also ongoing, among most groups with 13 different products being produced from nearly 43 tons of grain by 622 group members, mostly women. The groups also identified their most severe production constraints for maize, soybeans and beans, and recognized widespread plant nutrient deficiencies of nitrogen and phosphorus.

The N2Africa Project strongly influences the groups as well, promoting BNF technologies and encouraging groups to establish farm input shops, collective marketing centres and value-addition of grain legumes. Awareness of bean disorders and soil constraints was advanced through the recently established NIFA-Better Beans field campaign. The level that the special interests of both women and youth are represented at among these groups is impressive, and suggests that new project activities advancing their interests will receive ready collaboration through WeRATE. All WeRATE members requested support for additional farmer training, particularly in new farm technologies, 62 per cent in marketing and 48 per cent in agri-business (Woomer *et al.*, 2014).

New approach: the Agricultural Technology Clearinghouse

It is only fair that, when projects engage WeRATE for multi-site technology testing and popularization of new farm technologies, they also consider the stated needs of the NGO and its members. As a result, in response to growing

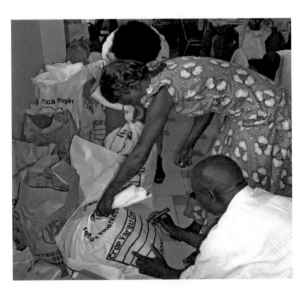

Figure 6.3 Participants at the WeRATE Planning Workshop and Second Agricultural Technology Clearinghouse for the 2015 long rains assembling test kits

Photo: WeRATE

interest in WeRATE coordination of technology testing and field campaigns, the seasonal Agricultural Technology Clearinghouse was organized. The Clearinghouse brings proven new farm technologies to its members by first introducing a suite of R4D projects and their field protocols and then soliciting member participation. This approach leads to specific agreements between WeRATE, its projects and members, and the logistics needed to deploy these field tests.

The first Clearinghouse was conducted in over three days in preparation for the 2014–2015 short rains growing season and attended by platform stakeholders. The Second Agricultural Technology Clearinghouse for the 2015 long rains took place in late February (Figures 6.1 and 6.3). During this workshop, WeRATE members were introduced to four different technology tests and provided opportunity to explore their usefulness during the short rains. These technologies (and projects) included BNF technologies (N2Africa), striga elimination (Humidtropics Action Research), better bean production (NIFA-Black Carbon) and cassava management (IFAD-Cassava). A short description of each technology test follows.

N2Africa BNF Best Practice

This test examines N2Africa Best Practice of mineral fertilization with Sympal and inoculation with BIOFIX, purchased from MEA Fertilizers Ltd, on soybean cv Squire, the best performing variety from last season's variety test provided by the Kenya Soybean Farmer Association. In all, 25 input packages were assembled and distributed to participants. Data report forms were submitted by 22 subscribers, an 88 per cent response. Results from this trial (Table 6.1) indicate that the recommended N2Africa package increases soybean yield by +860 kg/ha in part due to better plant stand and symbiotic performance. Subscribers to this trial demonstrated their ability to assess legume root nodules by several criteria. The next planned action is to evaluate the rate of Sympal application in different soils and agro-ecological zones of West Kenya.

Table 6.1 Soybean cv Squire yield, stand and nodulation characteristics in response to management on 22 farms in West Kenya during the 2014–2015 short rains growing season (± SEM)

Management	Grain yield (t/ha)	Plant stand (%)	Root nodules/ plant	crown nodulation (%)	Red interior (%)
No inputs	1.36 ± 0.22	80 ± 9	8 ± 2	3 ± 2	58 ± 13
Sympal[a]	1.65 ± 0.33	81 ± 8	11 ± 2	15 ± 8	65 ± 11
BIOFIX[b]	1.73 ± 0.23	83 ± 9	18 ± 2	47 ± 13	79 ± 10
Both inputs	2.22 ± 0.40	86 ± 9	27 ± 3	70 ± 11	85 ± 9

[a] Sympal Fertilizer blend (0–23–16+) at 125 kg/ha.
[b] BIOFIX legume inoculant (strain USDA 110) applied at 10g/kg of seed.

Source: WeRATE research

NIFA-Better Beans

This test involves beans and the benefits from better management, including the use of biochar as a soil amendment. There are ten managements in this test, the most complex evaluation WeRATE has undertaken. In all, 20 Better Beans technology packages were distributed to the leaders of farmer associations in West Kenya. Assembly of these packages was complex. NIFA provided about 500 kg of biochar packed in 7 kg bags, but they were very leaky, therefore WeRATE provided tightly woven polythene bags for repackaging. CIAT provided 80 kg of New Rose Coco (bush) seed of excellent quality. Options for climbing bean cv Tamu were available as well, with only four leaders selecting the latter. Inoculant packets (10 g) and fertilizers (1 kg) were specially packed by, and purchased from the MEA factory in Nakuru. Data reports were received from all 20 subscribers, but some responses were incomplete.

Preliminary results from these on-farm tests (Table 6.2) suggests that the recommended N2Africa technology package performs well (+314 kg/ha), is further enhanced through the addition of biochar (+134 kg/ha), due in part through modest disease suppression, and is greatest when mineral nitrogen is also applied (+136 kg/ha). The economic response to biochar is uncertain, however, as no commercial stocks are available so it remains difficult to price this experimental input. Subscribers not only demonstrated an ability to assess yield and nodulation, but also ranked severity of pests and disease (Figure 6.4).

Figure 6.4 WeRATE members identify preferred soybean management system – rust tolerant cv Squire variety – during farm liaison training

Photo: WeRATE

Table 6.2 Summarized results from the Better Bean trials on 20 farms show strong response to inputs and reduced root rot

Management	Bean yield (kg/ha)	Nodules/plant	Root rot (0–3 ranking)
No inputs	829±219	7±3	1.19±0.16
N2Africa package[a]	1143±304	19±4	0.88±0.25
Package with biochar[b]	1277±236	20±4	0.61±0.25
With biochar and CAN[c]	1413±254	16±4	0.73±0.24

[a] N2Africa package = Sympal fertilizer (276 kg/ha) and BIOFIX inoculant (USDA 2667).
[b] biochar applied at 2 t/ha.
[c] CAN (63 kg N/ha) replaces BIOFIX in N2Africa package.

Source: WeRATE research

Despite the sound performance of WeRATE subscribers, NIFA scientists elected to discontinue our collaboration after only one season. Instead they entered into direct agreement with individual WeRATE members they met through a field tour organized by the Platform. However, WeRATE members are developing their own technology tests under Better Beans II activity in 2015 long rains.

Humidtropics maize technologies

This test is designed to diagnose the severity of striga infestation and Maize Lethal Necrosis Virus (MLNV), and to evaluate the resistance of six newly released varieties from three commercial seed companies (Freshco, SeedCo and Western Seed Co.). A known susceptible maize variety (WH 403) serves as a control management and a sorghum–soybean intercrop offers an alternative to maize in the worst affected areas. This test includes the new imazapyr-resistant (IR) maize variety FRS 425-C. In all, 25 test packages were provided to WeRATE members for testing in striga and MLNV-infested areas.

All six managements receive a basal application of DAP and later CAN topdressing, inputs pre-packaged by MEA Ltd. The sorghum variety is a dwarf white type with a large market demand and its soybean intercrop is inoculated cv Squire. Data report forms on these on-farm tests were returned by 16 subscribers (88 per cent response); findings appear in Table 6.3.

IR maize performed well in striga-infested areas (Figure 6.5) and WH 402 expresses impressive tolerance to MLNV. The two highly productive hybrids (WH 507 and SC Simba) have reduced capacity to withstand these constraints, suggesting that farmers in infested areas are better advised to choose their maize varieties on the basis of specific tolerance rather than general yield potential. The sorghum–soybean intercrop tolerates striga and avoids MLNV but offers reduced yields, in part through reduced plant stands; results are currently undergoing economic analysis. Subscribers demonstrated their abilities to collect data directly related to two severe biotic constraints of maize but it is

Table 6.3 Performance of maize varieties and non-host intercrop in striga and/or
         MLNV-infested fields of West Kenya during the 2014–2015 short rains based
         upon 16 on-farm technology tests (±SEM)

Management	Strategy	Crop stand (plant/seed)	Crop yield (t/ha)	Striga stems per plant	MLNV tolerance (0–1 rank)
WH 403	Susceptible variety	0.89±0.12	1.95±0.36	5.4±1.8	0.5
FRC 425 IR[a]	Striga elimination by IR	0.93±0.14	3.02±0.39	3.1±1.6	0.4
WH 402	MLNV manage-ment	0.91±0.14	3.03±0.40	6.0±1.8	0.6
WH 507	Outgrow biotic stress	0.91±0.15	2.14±0.31	5.8±1.4	0.4
SC Simba	Outgrow biotic stress	0.85±0.14	2.45±0.36	5.6±2.0	0.5
Sila/Squire[b]	Non-host inter-cropping	0.75±0.23	1.08±0.28	2.1±0.4	0

[a] IR = Imazapyr resistant maize.
[b] Alternate rows of sorghum cv Sila and soybean cv Squire. All others are commercial maize
    varieties available in Kenya.

Source: WeRATE research

important that future field sites be more carefully selected for the presence and
degree of field infestation. Training will be offered to Master Farmers in this
regard. In the 2015 long rains, tests will be designed to diagnose the severity
of striga infestation and MLNV, and to evaluate the resistance of six newly
released varieties from the three commercial seed companies.

IFAD-Cassava

This test examines the effects of improving cassava variety, mineral fertilization, spacing and intercropping within eight different managements. In all, 18
cassava technology test kits and three cassava bulking packages were assembled
and assigned to WeRATE members. About 3,500 good quality cuttings from
a common (cv Merry Kalore), improved, released (cv Migera) and four KARI
experimental varieties (MM 96, 97, 98 and TR 14) were obtained from
pioneering efforts in Migori County. One of these cassava varieties, Migera, is
known for its leaf quality and over half the Master Farmers were familiar with
cassava leaf used as vegetable. Odd lots of these cassava cuttings were also
provided to members along with fertilizer for planting and multiplication. Early
assessment of these varieties is underway, in part using participatory methods
led by two graduate students from Masinde Muliro University of Science and
Technology. Recognized opportunities for improved cassava production have
opened doors to three county extension offices (Bungoma, Busia and Migori
Counties), collaboration that was previously difficult to forge. Next efforts will

Figure 6.5 WeRATE striga management approaches and farmer response: maize is
overwhelmed by intense striga infestation (left) that is greatly reduced by IR
maize (centre). Farmers synthesize field experience to develop a practical,
inexpensive strip-crop approach to striga elimination (right)

Photos: WeRATE

focus upon establishing legume understories within cassava production areas
(Obiero, 2014).

Clearinghouse assessment

In all, 88 technology packages, field protocols and data report forms were
distributed to 24 grass-roots organizations within the West Kenya Action Site.
This combined action also led to 27 farmer field days in conjunction with local
agricultural extension, farm input distributors and schools. Our Clearinghouse
approach proved particularly effective because in the past each project held its
own separate meetings and there was little coordination between them in terms
of input assembly, site selection, deployment and farmer field days. The
Clearinghouse process and participants are more fully described in a report
prepared by WeRATE (2014).

Operating within the Humidtropics research landscape

It is perhaps one advantage for an umbrella NGO such as WeRATE to operate
effectively on behalf of its members, and another to serve as a complete R4D
Platform that also assists CGIAR scientists to undertake difficult developmental
research tasks. At the same time, some interests are parallel, such as how to best
scale up a promising new technology, while others are tangential, such as
monitoring and interpreting farming system trade-off, or interpreting impacts

at a range of scales. In terms of scaling up new crop varieties and farm technologies, WeRATE and its partners have demonstrated considerable success in the areas of imazapyr-resistant maize to combat striga, introduction of improved climbing bean and soybean varieties, and creating demand for BNF technologies, particularly BIOFIX legume inoculants and Sympal blended fertilizer. Crop variety assessment is forwarded through arrangement for pre-release agreements so that farmers become familiar with new crop varieties as they also undergo Kenya's rather lengthy certification and release process.

Working with CIMMYT, KARI and AATF, WeRATE introduced IR maize to tens of thousands of households, creating a massive demand once the product reached stockist's shelves (Woomer *et al.*, 2008). Just as BIOFIX inoculant was licensed by the University of Nairobi to MEA Fertilizers Ltd, the N2Africa Project enlisted WeRATE to field test legume inoculation (Table 6.4), helping to create demand that resulted in an annual threefold increased inoculant production between 2010 and 2013. The development of Sympal fertilizer blend resulted from an even closer relationship because the product resulted from formulation, field testing and refinement by WeRATE and its partners, and within three years hundreds of tons of this blend were reaching farmers through commercial channels. Starting with only 650 kg of improved soybean seed in early 2010, WeRATE farmers reported over 6,000 tons of production after six seasons (three years). Over four years in collaboration with the N2Africa Project, WeRATE members reached over 37,000 households with a 64 per cent adoption of its best practice soybean variety–inoculant–fertilizer blend technology. WeRATE groups not only test and promote new farm products, but also establish their own farm input supply shops that offer 'last-mile' product delivery and offer discounts to members. Systems trade-offs are more difficult to track.

Scientists seek help from R4D Platforms to better understand which trade-offs occur and how these maximize farm production and yield. Within the Western Kenya Action Site, trade-offs occur through the greater recognition and understanding of both chronic and emerging challenges to production, as

Table 6.4 Summary of WeRATE outreach activities in West Kenya over four years (2010–13) through partnership with the N2Africa Project

Outreach action	Total
Number of new households	37,464
Number of on-farm demonstrations	355
Inoculant packets distributed	59,231
Legume seed distributed	223 tons
Fertilizer distributed (tons)	320 tons
Master farmers trained	226
Extension manuals distributed	48,938

Source: WeRATE research

Table 6.5 Nutritional composition of soymilk vs cow milk

Constituents	Soymilk (%)	Cow milk (%)
Proteins	5.7	3.5
Lipids	2.4	4.0
Carbohydrates	1.4	4.2
Minerals	0.8	0.7
Water	90	88

Source: Mulei *et al.* (2011, p. 8)

well as changes in market opportunities. Most farms practise maize–bean intercropping, and the invasion by striga and plant diseases have forced farmers to change their traditional crops and practices. Farmers belonging to one WeRATE founding member (MFAGRO in Vihiga) were the first in Africa to eradicate striga by adopting new control practices and blending them into acceptable community practice (AATF, 2006). Invasion of MLNV into new areas forces farmers to change crops, and WeRATE has sensitized the farming community to the threat and appropriate response to this rapidly spreading virus disease.

Trade-offs also occur among households adopting climbing beans and soybean. Climbing beans require support and several innovative staking systems have appeared. Soybeans were first intended for processors in urban markets but over time strong and more accessible local markets have emerged including buyers engaged in more localized processing and homemakers that better understand the nutritional advantages of this crop (Table 6.5). Even with its available detailed farming systems baseline, WeRATE is not well equipped to conduct complex trade-off analyses, but is a potential willing partner to scientists that step forward with resources, work plan and technical backstopping to do so.

WeRATE works on multiple crops but those of greatest interest to its members are maize, sorghum, beans, soybean and more recently root crops (cassava and sweet potato). Within the present scope of activities, WeRATE is able to simultaneously work on a wide range of field crops because of its participatory approach where individual member groups subscribe to different seasonal Clearinghouse activities. Admittedly, studies involving natural resource management, trees or livestock are longer term and require a different partici-patory structure.

Lessons learned and way forward

A large advantage exists in working with an umbrella NGO operating as an IP. It serves as a local coordinator for simple on-farm technology testing, farmer training and impact assessment and as a local partner for more complex research investigations. Its direct links to large numbers of farmers offers an alternative

WeRATE PLATFORM
(KENYA)

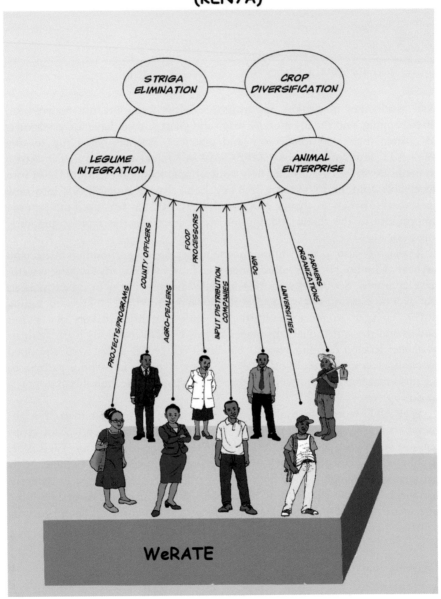

extension mechanism, especially where formal agricultural extension is weak. An umbrella structure allows for member groups to subscribe to specific opportunities of interest through an Agricultural Technology Clearinghouse approach.

Empowering an effective IP requires time and resources. Many members are unable to develop their own field campaigns and extension materials. In the case of umbrella organizations, officers of member groups are often unable to serve in a second, larger capacity requiring that the Platform appoint its own officers. Financial operations are challenging as the Platform must receive funds in a timely manner, distribute them to member groups according to specific agreement and assemble statements to acceptable standard. Some grass-roots WeRATE members, including those reliably submitting data and conducting dynamic field days, find it difficult to report finances to CGIAR standards, resulting in delayed release of funds the following season or year. Indeed, recognizing Platform shortcomings and developing incentives and training around them is a continuous process.

Real progress is made in improving productivity of maize–legume cropping systems but the individual households remain locked into poverty. The inputs required for improved production, such as IR-maize, specific fertilizers, legume inoculants, are known and available through agro-dealer channels, but poor households cannot afford them. Conducting technology demonstrations and farmer field days, and highlighting the achievements of early innovators is not sufficient for widespread impacts, and WeRATE and its partners must now become engaged in more innovative and better funded outreach. Value-added processing is critical to raising living standards in the smallest farms and this promising trend is noted among WeRATE members.

WeRATE was only recently formalized, and has not yet fully engaged in alliance with others, including the recently established county extension services. Previously, agricultural extension was managed at the national level, but constitutional changes have 'devolved' this responsibility to the counties. There are seven counties where WeRATE operates, it must better understand the different county rural development plans and find means to operate within them. On the other hand, WeRATE members have to establish strong linkage with the commercial sector, both farm input distributors and commodity buyers.

After a lengthy approval process by the NGO Board of Kenya, WeRATE now has its own bank accounts (both US$ and KES) and a KRA Pin Number. As a result, it will no longer manage funds through member accounts. This should lead to better financial reporting to supporters.

Challenges

The WeRATE R4D Platform has demonstrated its ability to conduct on-farm technology testing with a variety of research partners. Its Master Farmers have collected useful findings on crop yield, legume root nodulation, pest and diseases, and crop varietal comparison. Initially, some research partners were not in tune with the operations of the Platforms, in part because of expectations of excessive data collection and an unclear division between their project's research and outreach objectives. These differences were resolved through dialogue and development of mutually agreed field protocols. One challenge is to rectify the intention of some research projects to dictate where specific technologies are to be tested, and micromanage participation and incentives in a way that is potentially divisive to the Platform as a whole. For an innovative partnership to operate most effectively, a Platform must be seen as the leader of technology outreach, not inexpensive field labour. Indeed, WeRATE is operating along principles and with partners that permits this pioneering IP to advance proven technologies and new research products to their intended beneficiaries, Kenya's small scale farmers!

Appendices

See pages 113–115.

Appendix 6.1 WeRATE members, their areas of operation, farmer representation and group activities

WeRATE member	County	Email address	Reach (households)	Women chapter	Youth chapter	Liaison officers	Input shop	Seed production	Market produce	Process legumes
ARDAP	Busia	bonomondi2007@yahoo.com	10000	1	1	5	0	1	1	1
AVENE	Vihiga	avenecomdev@yahoo.com	1800	1	1	6	1	1	0	1
BUFFSO	Busia	livingstoneosuru@yahoo.com	1600	1	1	9	1	1	1	1
BUSCO	Kakamega	dorcasakeyo@yahoo.com	1200	1	1	9	0	1	1	1
BUSOFA	Bungoma	jothammandila@yahoo.com	1270	1	1	5	1	1	1	0
BUSSFFO	Bungoma	bussffo@yahoo.com	1230	1	1	4	1	1	1	1
HAGONGLO	Siaya	magagalex@yahoo.com	400	1	1	6	1	1	1	1
HECOP	Kisumu	pkisimba@yahoo.com	1500	1	1	5	1	0	0	1
KENAF	Kakamega	etemesibrian@yahoo.com	4000	1	1	4	0	1	1	0
KHG	Kakamega	josecongoma@yahoo.com	350	1	1	4	1	1	1	1
KUFGO	Migori	wnyangaria@gmail.com	21	0	0	1	0	1	1	0
MFAGRO	Vihiga	mfagrofarmers@gmail.com	800	1	1	5	1	1	1	1
MUDIFESOF	Kakamega	mumiassoya@yahoo.com	1500	1	1	6	1	1	1	1
MUUNGANO	Busia	muunganodg@yahoo.com	600	1	1	5	1	1	1	1
OWDF	Busia	owdf20107@hotmail.com	4680	1	1	2	1	1	1	1
ROP	Kakamega	drsanjawa@gmail.com	30000	1	1	6	0	1	1	1
RPK	Vihiga	kalehah@gmail.com	300	1	1	5	0	1	1	1
SCC–VI	Siaya	paul.wabomba@yahoo.com	2850	1	1	5	0	1	1	0
SCODP	Siaya	scodp2012@gmail.com	4320	1	1	7	1	1	1	0
UCRC	Siaya	rachel.adipo@gmail.com	4000	1	1	3	0	1	1	0
KESOFA	Migori	kesofasoya@yahoo.com	6000	1	1	12	0	1	1	1
Total leverage			78421	0.95	0.95	114	12	0.95	0.90	0.71

Appendix 6.2 WeRATE member interviews conducted
17–20 February, 2015

WeRATE member interviews conducted on 17–20 February, 2015
Interviewed by Renee Bullock, IITA Gender Specialist

1   **Interviewee Name: Boniface Omondi – ARDAP**
   a)   How has WeRATE helped you or your work?
       (i)   WeRATE has helped farmers gain access to new technologies.
              It has linked research institutions and farmers. For example, new
              germplasms have been used.
       (ii)   WeRATE has helped build capacity by enabling farmers to
              understand technology dissemination and productivity.
   b)   How could WeRATE be improved?
       (i)   Since farmer involvement is key, a participatory approach is
              needed. We could help farmers to understand the process, such
              as identifying problems and working together.
       (ii)   Sometimes they do not understand interventions that are
              developed and why they are brought to them.
       (iii) The platform could link local organizations to input suppliers and
              larger input distributers.

2   **Interviewee Name: Pam Ogutu – HAGONGLO**
   a)   How has WeRATE helped you or your work?
       (i)   WeRATE has helped in facilitation, training and dissemination
              of how to use technologies and value addition.
   b)   How could WeRATE be improved?
       (i)   More trainings are needed to reach farmers, we should find ways
              to reach larger areas.
       (ii)   We should develop more technologies on different crops, i.e.
              diversification.
       (iii) We need more gender action and to work together with youth
              to make a difference.

3   **Interviewee Name: Dorcas Akeyo – BUSCO**
   a)   How has WeRATE helped you or your work?
       (i)   Value addition such as processing has empowered women to earn
              money from products that include milk, flour, and crunchies.
       (ii)   We sell grains to companies in Nairobi and villages.
   b)   How could WeRATE be improved?
       (i)   We need to empower women and youth by encouraging value
              addition.

4   **Interviewee Name: John Onyango – KESOFA**
   a)   How has WeRATE helped you or your work?
   b)   How could WeRATE be improved?

(i)   We need to strengthen governance of the platform. During elections there is a need to pull from different regions so they are all included and not any one area is favoured.

(ii)   In management we should create a position like a programme manager to report to. That one person manages others and reports to Project Coordinators.

5   Interviewee Name: Rachel Adipo – UCRC

a)   How has WeRATE helped you or your work?

(i)   We have benefited from soya. Prices of soya used to be very high and therefore unaffordable. Now the prices are lower and more people can buy them.

(ii)   Marketing links have been created between farmers and the platform.

(iii)   Field days increase awareness.

(iv)   Household nutrition and soil fertility have improved.

b)   How could WeRATE be improved?

(i)   Communication could be improved. Rachel would like to be directly contacted and would like more communication with members in her organization so they realize the importance of WeRATE activities.

Appendix 6.3 Agro-ecological zones in the West Kenya action site

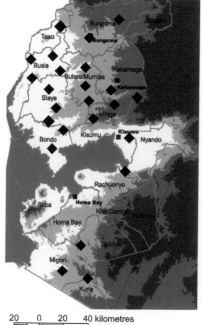

Three major agro-ecological zones occur in WeRATE's Action Area

Lake Victoria Basin (1125–1300 masl): semi-arid to semi-humid climate, maize-based cropping with some cassava and rice. Failing cotton.

Lower Midlands (1300–1500 masl): sub-humid climate with rolling hills and plateaus, maize–bean intercropping with sweet potato, banana. Large sugar plantations and out-growers. Failing tobacco.

Upper Midlands (1500–1800 masl): humid climate, mountainous terrain, maize–bean cropping with potato, pea and vegetables. Tea out-growers.

Altitude in meters		
< – 1300	Lake Basin	◆ WeRATE
1301–1500	Midlands	Site
1501–1800	Upper Midlands	
> – 1800	Highlands	

20   0   20   40 kilometres

References

African Agricultural Technology Foundation, 2006. Empowering African Farmers to Eradicate Striga from Maize Croplands. A call to action by The African Agricultural Technology Foundation (AATF), Nairobi, Kenya.

Humidtropics, 2012. Humidtropics: Integrated Systems for the Humid Tropics (Project Document). International Institute of Tropical Agriculture (IITA), Ibadan, Nigeria.

Mulei, W.M., Ibumi, M., Woomer, P.L., 2011. *Grain Legume Processing Handbook: Value Addition to Bean, Cowpea, Groundnut and Soybean by Small-Scale African Farmers.* Tropical Soil Biology and Fertility Institute of the International Center for Tropical Agriculture (CIAT), Nairobi, Kenya.

Obiero, H., 2014. IFAD-Cassava Project Western Kenya, 2014 Report. WeRATE, Mbale, Kenya.

Okalebo, J.R., Othieno, C.O., Woomer, P.L., Karanja, N.K., Semoka, J.R.M., Bekunda, M.A., Mugendi, D.N., Muasya, R.M., Bationo, A., Mukhwana, E.J., 2006. Available technologies to replenish soil fertility in East Africa. *Nutrient Recycling in Agroecosystems* 76, 153–170.

Sanginga, N., Woomer, P.L. (eds), 2009. *Integrated Soil Fertility Management in Africa: Principles, Practices and Developmental Process.* Tropical Soil Biology and Fertility Institute of the International Center for Tropical Agriculture (CIAT), Nairobi, Kenya.

Savala, C.E.N., Omare, M.N., Woomer, P.L. (eds), 2003. *Organic Resource Management in Kenya: Perspectives and Guidelines.* Forum for Organic Resource Management and Agricultural Technologies (FORMAT), Nairobi, Kenya.

WeRATE, 2014. *WeRATE Planning Workshop Report for the 2014–2015 Short Rains: A new technology clearinghouse approach.* WeRATE, Mbale, Kenya.

Woomer, P.L., 2007. Costs and returns to soil fertility management options in Western Kenya. In: Bationo, A. (ed.), *Advances in Integrated Soil Fertility Research in Sub-Saharan Africa: Challenges and Opportunities.* Springer Scientific Publishers, Midrand, South Africa, 877–885.

Woomer, P.L., Mukhwana, E.J., Lynam, J.K., 2002. On-Farm research and operational strategies in soil fertility management. In: Vanlauwe, B., Diels, J., Sanginga, N., Merckx, R. (eds), *Integrated Plant Nutrient Management in Sub-Saharan Africa.* CABI, Wallingford, UK, 313–332.

Woomer, P.L., Bokanga, M., Odhiambo, G.D., 2008. *Striga* management and the African farmer. *Outlook on Agriculture* 37, 277–282.

Woomer, P.L., Karanja, N.K., Kaleha, C., 2014. N2Africa Kenya Country Annual Report for 2014: The challenging transition to Tier 1 status while maintaining strong project momentum. N2Africa Project, International Institute of Tropical Agriculture (IITA), Nairobi, Kenya.

7 Innovation platforms for improved natural resource management and sustainable intensification in the Ethiopian Highlands

Zelalem Lema, Annet Abenakyo Mulema, Ewen Le Borgne and Alan Duncan

System trade-offs call for IPs

The Ethiopian Highlands are a land degradation hotspot. The burgeoning human population has led to expansion of arable land to meet growing food demands. Much of this expansion is on steep and marginal land covered with fragile soils. The result has been extensive soil loss, sedimentation of water-courses and general land degradation that has affected production and the productivity of smallholder farmers. Addressing this problem requires both upstream and downstream land users, together with other kinds of people interested in the issues to work together and introduce interventions such as soil and water conservation structures. However, for farmers to invest in such structures, they need to provide financial benefits. Therefore, improved crop and livestock productivity, and marketing, need to feature. This led to the setting up of IPs to stimulate ongoing discussions among different kinds of people interested in the issues around natural resource management.

Three IPs[1] were set up under the Nile Basin Development Challenge (NBDC) to focus on implementing improved rainwater management practices to enhance the natural resource base for existing farming systems. These platforms focused on system integration in contrast to most other platforms that focus on a single commodity. They have proved to be effective in eliciting the kind of collective action at community and cross-sectoral level that is needed to positively stimulate sustainable intensification.

Initiation of the three platforms

Farming systems in the Ethiopian Highlands are characterized by mixed crop–livestock farming with complex problems that smallholder farmers and local development partners face. A large proportion of production is subsistence-

Figure 7.1 Jeldu District facing serious soil erosion
Photos: ILRI/Z. Lema

oriented which leaves farmers with limited capital to invest in interventions aimed at long-term improvements in productivity, such as soil and water conservation structures. One route to reversing land degradation is to intensify existing production of staple food crops, cash crops, livestock and trees. This will reduce the need to expand into land that is unsuitable for cultivation and at the same time generate the capital and financial incentives for farmers to invest in the land. Such intensification requires an integrated approach that takes into account the synergies and trade-offs between different farm enterprises. IPs provided the forum to discuss and experiment with intensification of multiple commodities and the approach has shown some early promise.

Evidence generated by Ludi *et al.* (2013) in the same districts stressed the need for multi-stakeholder processes to deal with rainwater management issues. The research showed that local stakeholders were expected to deliver on top-down targets and that difficulties were experienced in engaging farmers in planning and implementation. Farmers are more concerned about short-term incentives that increase the availability of food for their families and livestock feed in order to invest in land. That is why the current government of Ethiopia has struggled to build the required level of involvement from farmers. Farmers are facing constraints to feed their hungry livestock and most of them use free grazing which jeopardizes the sustainability of the local government initiatives on soil and water conservation structures. One of the approaches to implement integrated natural resource management (NRM) is through setting up IPs to provide space for relevant actors to jointly identify constraints and solutions to NRM issues at the local level (Nederlof and Pyburn, 2012).

ILRI Researchers working under the NBDC supported the establishment of three local IPs at the district (woreda[2]) level early in 2011. Platforms were established in three woredas, namely Jeldu, Fogera and Diga. After three years, two out of three platforms were adopted by the Humidtropics programme in 2014. The platforms have gone through a series of processes and stages that address the key constraints that farmers in the Ethiopian highlands face.

Figure 7.2 Diga woreda IP members
Photo: IWMI/D. Tadesse

Function of the platforms

At the initial stage, local stakeholders were identified as platform members based on their direct or indirect role in planning and implementation of NRM activities in each woreda. The majority of members are from the local government offices at woreda level and others include non-governmental organizations (NGOs), research centres, farmers and community leaders. Each platform has up to 30 members who have agreed to meet three to four times a year at the woreda headquarters to co-learn and coordinate joint activities.

In 2011, the platforms passed through a series of engagement activities to accommodate the various interests of its members. These activities included, but were not limited to: community engagement exercises and regular platform meetings organized to exchange knowledge to help members of the platform make informed decisions. After constructive dialogues among platform members, consensus was reached in identifying three key site-specific natural resource management issues:

- soil erosion for Jeldu;
- land degradation for Diga;
- free grazing for Fogera woreda.

These issues were highlighted by members as the key constraints to NRM that they wanted to address jointly as a priority. The members narrowed down on a specific intervention: improved and multipurpose livestock feed, which could be rolled out and tested in farmer's fields across all three locations. This intervention of improving livestock feed had great potential not only to address the problem of feed shortage, but also to boost soil and water conservation, thereby leading to more efficient natural resource management overall.

During the planning meeting, each platform developed its own working modalities to support the implementation of the interventions on livestock feed. They agreed to evaluate their interventions each year through actively participating in their regular meetings and visits to sites during farmer field days. For technical backstopping they selected members to form a technical group (TG) that represented key stakeholders to facilitate the meetings, implement the interventions and organize field days and exchange visits. Out of the 30 members, eight were selected based on the criteria that members had set, i.e. required multi-disciplinarity in the group and representativeness of organizations that have potential to implement the pilot interventions at scale. The TG members are similar across the three platforms and include technical staff from key organizations that have the potential to run the implementation. The members also agreed to follow up the progress of implementation through presentations of activities by TG members during their regular meetings. ILRI supported these TG members in each platform through backstopping and building of local capacity on forage interventions, and as much as possible devolved the leadership role to them.

Innovation fund to support fodder development

In 2012 and 2013 an innovation fund was established by ILRI as 'seed' money to support each platform's action on fodder development. The seed money was provided on the basis of proposals developed jointly by members, to enable piloting of their new approaches with participating farmers. The criteria for providing the seed money were that proposals had to be cross-sectoral, participatory, targeted at addressing local community concerns and scalable. The seed money was planned and used only to buy inputs, transport them to farmers' fields and support the trainings for farmers three times a year. The practical trainings were provided to participating farmers on their field during planting, management and utilization. Attention was given to developing farmers' capacity to harvest seed and seedlings and to expand plantation of the new fodder varieties with model farmers.

Farmers' interest in participating in improved livestock feed development arose because of its potential to address their pressing need of feed to feed their hungry livestock. Community engagement in problem identification, planning and implementation up to demonstration were central to the fodder inter-vention. Field-level trainings for farmers helped them to be able to plant, manage and utilize the feed resources efficiently. The trained farmers have the technical skills to collect seeds and seedlings before harvesting *Rhodes* grass and transplanting *Desho* grass seedlings and the practice is expanding to other farmers.

The role of different actors in scaling up

There is interest at district level to take innovations that work for farmers to scale. The key potential organizations to aid this are represented in each platform and engaged in the process of innovation generation, testing implementa-tion and monitoring so that they prove what works well among farmers. The evidence makes it easier for projects and government experts to expand the interventions at scale. Government projects and NGOs working on soil and water conservation have the financial capacity but lack inputs and technical capacity to fill some of their gaps including forage seed shortages. Involvement of district administrators in the regular learning meetings and farmers' field days (to see the feed interventions first hand) was found to be a good approach as it enabled them to realize that shortage of forage seed can be resolved if they worked closely with the participating farmers. Recently, local government and NGO projects have started working closely with the model farmers who served as community seed producers for livestock feed in order to maximize impact. During the field visits, local government and NGOs were impressed by the achievements and organized another farmers' field that brought together a large number of farmers to learn from the farmers participating in the platforms.

Platform members realized the importance of working together both at cross-sectoral and farmers' level. Local universities and NGOs also created a good network mainly around integrating their activities and resources. For instance Wollega University technically supported the platform interventions and also provided seed for Rhodes grass from its livestock feed demonstration research site. NGOs started supporting and working with local government staff by providing transport for inputs. More importantly, farmer-to-farmer linkages were created to disseminate the introduced livestock forage seeds and seedlings through selling and buying, with advice on how to plant and manage the seeds.

The work of the IPs in Diga and Jeldu on integrating the natural resource activities with livestock feed also attracted other CGIAR centres working on the Humidtropics programme (one of the CGIAR Research Programs) in Ethiopia. The International Water Management Institute (IWMI), the International Potato Center (CIP) and the World Agroforestry Center (ICRAF), joined the IPs to continue supporting the platform members to address the main crop production and market problems, develop the livestock feed market and to continue working on natural resource management.

Outcomes and impact of the intervention

The total number of model farmers who directly participated in the IP interventions during the NBDC project period in the three sites was 259. The model farmers were able to showcase the ability to feed their livestock during dry spells while maintaining their soils and natural resources.

After NBDC was phased out the Humidtropics programme continued working in the Jeldu and Diga sites with the IPs and started building on feed interventions and addressing other problems that farmers were facing. The work focused more heavily on increasing production and productivity of the main crops (maize, teff, barley and wheat) through improved management practices and improved seeds, integrated with livestock feed and natural resource conservation initiatives. An improved variety of sweet potato was also introduced in Diga. CGIAR centres including IWMI, CIP and ILRI have been working on the integrated approach in 2014 with 135 farmers.

Three years after implementation of the NBDC project, a qualitative study was undertaken by ILRI between March and June 2014 to assess the impact of the IPs. The study captured perceptions of changes that key categories of actors had made over the project period using indicators of change in knowledge, attitudes, skills and practices (KASP) with regard to soil and water management. The indicators were developed by NBDC partners as part of a project 'outcome logic model' (OLM) (Figure 7.3).

Farmers adopt forages and increase their skill set leading to livelihood and environmental benefits

As a result of the IPs, all 20 male and female farmers interviewed have gained knowledge and skills in the use of multipurpose soil and water conservation

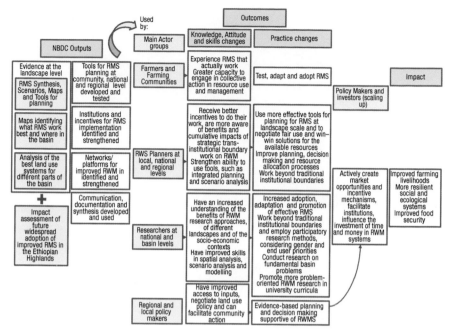

Figure 7.3 The NBDC programme outcome logic model
Source: Authors' research

(SWC) practices. These farmers have also applied the new practices introduced by the project, i.e. fodder development (using Sesbania, Chomo grass, Rhodes grass, Elephant grass and Desho grass), area closure, tree-planting and planting legumes (such as peas) to prevent soil erosion but also to provide feed for livestock (Figure 7.4). Although both male and female farmers apply similar SWC strategies, there are gendered preferences. Women mentioned planting grasses and legumes more frequently while men mentioned tree planting, terraces and bund construction more often. This could be attributed to the intensity of labour required for different practices and also the gendered farm practices where women are more associated with planting legumes such as peas and feeding animals kept at home while men are associated with the heavier tasks such as constructing dams and soil bunds.

Eleven male and seven female farmers across the three sites cited growing animal fodder and legumes, area closure and terrace construction as the most successful methods. Seventeen farmers (14 men and three women) identified forage development as the method that worked most effectively. They validated this by the benefits they have obtained in cash and in kind such as increased availability of alternative animal feed sources particularly during dry spells, increased crop productivity, regeneration of vegetation on previously degraded land, mitigation of termite damage, reduction in soil erosion and increase in milk yield and quality.

Figure 7.4 Farmers harvesting Desho grass and feeding their hungry animals

Photos: Tsehay Regassa (ILRI) (top), Zelalem Lema (ILRI) (bottom)

The five researchers, 24 planners and five policy makers interviewed also mentioned measurable benefits to farmers that have accrued from the adoption of the NRM practices such as reduction in soil erosion, less termite infestation and increased income from the sale of fodder. One unanticipated outcome was that farmers were able to sell seeds and thus gain income from the feed intervention. One kilogram of Rhodes grass seed goes for 150 Ethiopian Birr (ETB) (approx. 7 USD). In Fogera, for example, 9 tons of fodder were harvested from communal grazing land management (closed area), enabling 11 cattle to be fattened for market. In Diga more than 60 kg of Rhodes grass seeds were sold to government and NGO projects for scaling up. One male farmer named Leta in Diga planted Rhodes grass on one hectare of his private land for his fattening business. He bought four oxen for 4,000 ETB (approx.

Figure 7.5 Farmers in Jeldu harvest and feed Desho grasses planted on soil bunds
Photo: Desalegn Tadese (IWMI)

USD200) each and fed them Rhodes grass using the cut and carry system together with other complementary feeds for four months and sold each for 8,000 ETB (approx. USD400) (see Figure 7.5).

The farmers attributed the success of the practices to prior training by the NBDC/ILRI staff and other implementing partners, access to inputs, increased collaboration among stakeholders and cooperation at community and household level. A 40-year-old female farmer in Diga noted:

> Awareness has been created through the IP, follow-up and technical support by experts; farmers felt the need of forage development (which has multiple benefits for rehabilitating degraded land and managing termites) to supply livestock feed, and the potential source of income this brings adds to the reasons behind its success. Access to planting materials, fertilizer, technical backstopping, and farmers' commitment are also other major contributing factors.

In Diga most of the respondents have adopted fodder development, compost manure application and multipurpose tree species, while in Fogera terrace construction, area closure, fodder development and legumes are the methods most frequently adopted by farmers. Farmers in Jeldu have adopted all the SWC strategies that have complemented the government interventions. From the introduced fodder varieties, Desho grass (in Jeldu) and Rhodes and Chomo grass (in Diga) were chosen and taken up by a number of farmers while in Fogera

Figure 7.6 Farmers are demonstrating the different grasses during farmers' field day
Photos: IWMI/D. Tadesse (top) and ILRI/Z. Lema (bottom)

grazing land management has shown significant improvement to harvesting good biomass of natural grasses combined with legume fodder varieties. The number of farmers participating in fodder development has risen. In Jeldu for instance 96 farmers participated in the project intervention in 2012. This number rose to 141 in 2013.

Due to the knowledge and skills acquired, all the male and female farmers interviewed have made changes to the way they farm and have improved their farming practices from what they did three to five years back (Figure 7.6). This was at individual farm level as well as at the community level. For instance a female farmer in Diga explained that she has improved her skills in how to cultivate and manage improved forage over the last two years, and that her farm management skills have improved over the last year. She has learned how to make compost manure to use instead of fertilizer, which is too expensive. A 48-year-old man in Fogera previously used fertilizer under obligation but now he is happy to use fertilizer and other chemicals.

Increased collaboration among stakeholders

Farmers in all the sites felt that participation in the NBDC programme has increased their level of cooperation which has led to more effective management of soil and water in their areas. The project has developed farmers' capacity to carry out SWC activities better: two female and one male farmer in Fogera recounted that:

> Some three years back we were working separately on our private fields and directed erosion downstream, which was the cause of land degradation; but starting from 2012 we planned and implemented soil and water management practices with full participation of the community. The change has come because of the information given ahead of implementation.

Increased cooperation between husband and wife was also noted due to a change in constraining norms, negative attitudes and perceptions about women and their involvement in NRM. Indeed, farmers attributed the change in attitude and practices to the project, stating:

> Thanks to awareness creation by ILRI, women's involvement in NRM – particularly planting grasses – has increased. This is due to the awareness creation for both women and men by the government and ILRI that urged the necessity of collective work on NRM. As a result women and men's collaboration on soil and water management practices has increased.
>
> (Jeldu farmer, male)

Increase in collaboration was felt not only by farmers but also by planners, researchers and policy makers who now collaborate more strongly with other stakeholders than they did before the NBDC. Back then, institutions worked independently; where institutions did have partners, their level of engagement increased after being involved in the project. One policy maker recalled: 'Before, all institutions were working independently but now even government insists that we work with different institutions. It's a government policy even within Oromia region. This policy encourages working with farmers.'

As a result of exposure to integrated participatory planning tools, planners and researchers changed their approach of engaging with other stakeholders to

INNOVATION PLATFORM FOR IMPROVED NRM AND SUSTAINABLE INTENSIFICATION (ETHIOPIA)

INNOVATION PLATFORM FOR IMPROVED NRM AND SUSTAINABLE INTENSIFICATION (ETHIOPIA)

5

include not only experts but development agents as well as farmers. The actual problem–solving approach is based on listening to the voices of the voiceless, the local people using participatory and integrated planning tools. Soil and water management was based on community members identifying their problems and seeing themselves as part of the solution. Although integrated planning tools were used in the NBDC programme, their practical use and integration within the ongoing government NRM interventions was hindered by budgetary constraints and the stringent top-down government planning procedure.

Lessons learned

The key challenges faced during the implementation of improved fodder through IPs were time, incentives and not being able to realize outcomes in a short period of time. One of the time-consuming activities was prioritizing site-specific constraints which took one year because of differing interests among IP members. The platform was represented by the majority of public government line ministries that have their own targets to achieve and dominate/override the community's interest. It required strong facilitation skills to mediate and that is why ILRI's research team undertook community engagement exercises to get community members' interests in front of the platform for consideration. The other challenges were high staff turn-over and lack of consistent participation of IP members (a problem for building the local research capacity to innovate). Since all the stakeholders have other assignments, bringing them together and getting them to commit their time to IP activities on a voluntary basis is a big challenge.

Conclusion

The case of NBDC IPs nested into the Humidtropics programme highlights several important lessons. Setting up IPs significantly raised the knowledge of farmers involved in soil and water conservation practices, and farmers applied this knowledge effectively in their own practice. IPs also raised a collective sense of belonging, collaboration and collective action, across all the stakeholder groups involved (including farmers, planners, researchers and policy makers). Furthermore, the multi-stakeholder, multi-meeting, multi-year nature of IPs seems to have highlighted a much richer set of interlinked issues (e.g. soil erosion, climate change adaptation, termite degradation etc.) than the original focus (e.g. improving soil erosion and land degradation through feeds and forages). This helped all actors involved focus on the bigger picture and deal with it more systemically, while creating opportunities for further initiatives and interactions to deal with new issues coming up.

Another crucial lesson emerged for IPs that focus on natural resource management, where gains are typically obtained in the longer term. For such platforms, early economic wins (e.g. cattle fattening leading to more money) seem to be essential incentives, and arguably the main reason why farmers were happy to invest in better NRM – however, other incentives (recognition, capacity development) could also be important incentives in the medium to longer run. Despite these positive lessons, we must remember that aligning visions and agendas to identify the most crucial challenges collectively took a year. This raises questions about the ease of replicating IPs as a development approach.

Moving ahead

This case study sets out some important elements for consideration by other IPs dealing with NRM. The next frontier for this set of IPs, and with general research on IPs, relates to sustainability and scaling up. There are a few key questions to ponder in this regard. First, in the lifetime of a project using IPs, how can one foster collective capacity to innovate with limited inputs, time span and high turnover of personnel? What supplementary measures, aside from the specific work done at the platform meetings, can really enhance that capacity to innovate? Second, in a government-dominated state with top-down decision-making processes, what are the best options to institutionalize the participation, co-creation and innovation dynamics that IPs tend to bring about? And lastly, if state agencies are taken by the idea of bottom-up decision making, how can we ensure that IPs are not used for 'token participation', as is too often the case? We are hopeful that the small scale but successful NBDC platforms that are now addressing the Humidtropics challenge will unravel some of these puzzles soon.

Acknowledgements

This work was funded initially by the Challenge Programme for Water and Food (CPWF) and latterly by the CGIAR Humidtropics programme. Thanks to Gerba Leta who assisted with the impact study.

Notes

1   http://nilebdc.org/2011/06/30/local-innovation-platforms-for-the-nile-bdc/   – accessed 8 January 2015.
2   Woreda is the third level of administrative divisions in Ethiopia and it is the basic decentralized administrative unit managed by local governments.

References

Ludi, E., Belay, A., Duncan, A., Snyder, K., Tucker, J., Cullen, B., Belissa, M., Oljira, T., Teferi, A., Nigussie, Z., Deresse, A., Debela, M., Chanie, Y., Lule, D., Samuel, D., Lema, Z., Berhanu, A., Merrey, D.J., 2013. Rhetoric versus realities: A diagnosis of rainwater management development processes in the Blue Nile Basin of Ethiopia.

CPWF Research for Development (R4D) Series 5. CGIAR Challenge Program on Water and Food (CPWF), Colombo, Sri Lanka.

Nederlof, E.S., Pyburn, R. (eds), 2012. *One Finger Cannot Lift a Rock: Facilitating Innovation Platforms to Trigger Institutional Change in West-Africa.* KIT-Royal Tropical Institute, Amsterdam, the Netherlands.

8 Sustaining the supply of organic White Gold

The case of SysCom innovation platforms in India

Christian Andres, Lokendra Singh Mandloi and Gurbir Singh Bhullar

Setting the scene

White Gold: a primer

Why White Gold? Cotton (*Gossypium* spp.), also known as "White Gold," is not only the most important fiber plant for the production of textiles, but also one of the most intensive crops in terms of pesticide use worldwide (Bachmann, 2012). That's why the genetically modified *Bt* cotton was developed, which gives protection against the most important cotton pests: the bollworms (*Helicoverpa* spp.).

Let's go to India, the mother of history, the grandmother of legends. Madhya Pradesh State is located in the central cotton belt of dryland India. Here, *Bt* cotton occupies more than 90 percent of the cotton area (Choudhary, 2010). However, at the same time, it is also the biggest producer state of organic cotton (Truscott *et al.*, 2013).

Organic is better, isn't it?

Many of us would say yes, of course. These days, statements such as the following are springing up like mushrooms: "Increasing concerns about global food security, depleting fossil reserves and diminishing natural resources question the continuation of energy-intensive conventional agriculture, and emphasize the importance of sustainable alternatives such as organic agriculture" (IAASTD, 2009). But why would Mr. Manjit Singh Dang, an Indian small-scale farmer who produces the organic White Gold, choose organic over conventional? After all, the latter is not only less complicated, but often also more productive and thus more rewarding, right?

In Switzerland, the case is crystal clear. The Research Institute of Organic Agriculture (FiBL) has shown that organic farming leads to lower yields, but has many other benefits compared to conventional farming (Mäder *et al.*, 2002).

Today, Coop (the biggest retailer of organic products in Switzerland) makes a turnover of over one billion USD with organic products. When we go further south though, the picture becomes blurred. There is little scientific data on the comparative performance of organic vs. conventional farming systems in (sub)tropical zones. That's why FiBL launched a large program called Systems Comparisons in the Tropics (SysCom[1]). SysCom provides innovation platforms (IPs) in three countries: Bolivia, India and Kenya. It maintains a network of long-term farming systems comparison experiments (LTEs) and addresses specific challenges of small-scale organic farmers through participatory on-farm research (POR).

Besides cotton as his main cash crop, Manjit cultivates soybeans and wheat. But he had a problem: while his conventional colleagues were very flexible in terms of crop management strategies, his yields heavily relied on the limited options allowed in organic farming. Phosphorous (P) nutrition was a particular problem, because the local soils are highly alkaline; in fact, so alkaline that the usual organic P fertilizer (rock phosphate (RP)) did not work. So Manjit had no suitable phosphorous fertilizer, so both the yield and the fiber quality of his White Gold was low. This case study illustrates how Manjit and his fellow farmers overcame these limitations by being part of an IP.

How to get more organic White Gold?

Good question. We propose focusing on three central questions:

1   What can we do to increase the productivity of organic cotton systems on alkaline soils?
2   How can we spread innovations among small scale farmers efficiently?
3   How to increase the attractiveness of organic cotton systems?

Let us just sneak a peek of what is to come: through the IP in India, we developed a new kind of high quality phospho-compost that is produced from RP, butter milk and well-stored farmyard manure. Our farmers increased the yields of their White Gold and soybeans by 40 percent on average with this new technology. However, the most impactful thing we did to spread the innovation was to launch a competition among the participating farmers, arguably a more powerful tool for IPs to create impact than normal word of mouth strategies. And last but not least, we created scientific evidence that despite lower yields in organic cotton systems, the lower production costs rendered them equally rewarding as conventional systems. The less capital-intensive nature of organic cotton systems can have important implications when crops fail.

Madhya Pradesh's organic cotton problem and potential ways out

A vast problem lying in the valley of a holy river

"Narmada never runs dry you know, it is holy. A teardrop that fell from the eyes of Lord Brahma, the creator of the universe, yielded the river." Just one of the "legend has it" statements you'll hear from locals when you ask them about the many pumps and pipelines lining the shore of Narmada. Fact is that agriculture in the plains of the river heavily relies on its water for irrigation. Narmada has shaped the landscape, creating Vertisols (also known as "black cotton soils") that stretch approximately 5km to both sides of the river. These soils are mostly fertile, but also highly alkaline which poses a particular problem for crop nutrition in the production of organic White Gold.

As mentioned, Manjit had a problem with phosphorous: he used to apply RP which did not show any effect on his alkaline soils due to chemical processes (Appendix 8.1). Manjit was neither aware of that, nor did he have any other choice in terms of organic P fertilizer. Conventional farmers don't have this problem. They can use synthetic P fertilizers that work on alkaline soils. These fertilizers are produced by treating RP with strong inorganic acids.

The scope of this problem is vast: vertisols are not only the predominant soil type in Madhya Pradesh, but they cover a staggering 73 million hectares of the subtropical regions of India (Kanwar, 1988). The country counted 184,029 farmers producing 75 percent of the world's supply of organic cotton (312,131 Mt seed cotton) in 2011–12; 50 percent of this amount was produced by 90,500 small scale farmers in Madhya Pradesh (Truscott *et al.*, 2013). bioRe[®] India Ltd. is an organic cotton enterprise that works with some 5,000 small scale farmers (bioRe farmers). The company mainly operates in the Khargone district (area: 8,030 km[2]) of Madhya Pradesh (Figure 8.1). Other major Indian states growing the White Gold include Andhra Pradesh, Gujarat, Haryana, Karnataka, Maharashtra, Odisha, Punjab, Rajasthan and Tamil Nadu (Figure 8.1).

Fighting complexity with diversity

To address this rather complex problem, FiBL set up an IP at bioRe back in 2006. The IP brings together a wide range of stakeholders in order to ensure the acceptance of our activities at different levels. Among them are bioRe[®] India Ltd. and its farmers, an associated non-profit organization (bioRe Association[2]), and researchers from both India and Switzerland. Furthermore, an Indian spinning mill, a Swiss yarn trader (Remei AG), and donors representing NGOs (Biovision Foundation for Ecological Development), retailers (Coop Sustainability Fund) and governmental development agencies from Switzerland (Swiss Agency for Development and Cooperation) and Liechtenstein (Liechtenstein Development Service) were involved. Details about the stakeholders are provided in Appendix 8.2.

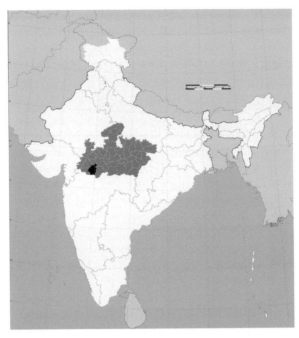

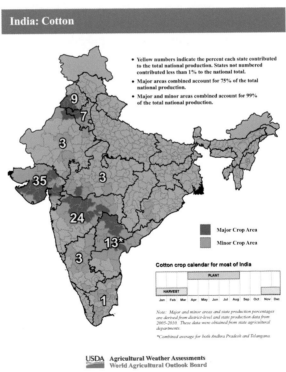

Figure 8.1 Top: location of Khargone District (black) in Madhya Pradesh State (red).
Bottom: major cotton growing areas of India

Source: Wikimedia (top) and Spectrum commodities (bottom)

The centerpiece of the IP is the long-term farming systems comparison experiment (LTE). The agronomic on-station experiment is carried out at the training and education center of bioRe Association, and has as its main objective to create scientific evidence about the comparative performance of organic vs. conventional cotton systems. In order to ensure that the LTE represents local farming systems, we meet twice a year with a Farmers Advisory Committee (consisting of five representatives of conventional and organic farmers each): one time to plan the season, and another time to evaluate the performance of the crops.

But creating evidence and papers is not enough, especially for farmers. After all, paper remains paper. Farmers want to see hands-on solutions for their problems from us researchers, and rather today than tomorrow. That's why we launched the participatory on-farm research (POR) component back in 2009. The goal of POR is to develop innovations that improve yields and rural livelihoods of local small-scale farmers in the mid to long term. In working with the farmers, we chose a combination of the LTE (e.g. for demonstration trials) and several POR trials (e.g. for exchange visits).

The birth of the RP-FYM technology

So how do you start such a process? First, we had to identify the needs of our beneficiaries. "So let's ask about the main challenges of our farmers," we thought. Nothing easier than that one could assume, but if you find yourself standing in front of 150 farmers it turns into a major challenge. So there we were: researchers and farmers in a ratio of 1:50, trying to reach a conclusion. We did semi-structured interviews and focus group discussions with Manjit and his colleagues (Figure 8.2). Finally, after asking countless questions and a prolonged discussion we reached consensus: together we wanted to work on the P problem described above.

It was clear: the efficiency of RP had to be improved so that the 5,000 bioRe farmers could enhance their yields, and the fiber quality of White Gold. We set off on our journey by identifying several local materials with a potential to solubilize RP (in order to make it easier for plants to absorb). Farmers and extension agents suggested compost, phosphorus solubilizing bacteria, tamarind fruits, local vinegar (LV) and buttermilk (BM). So we screened these materials in a first set of trials in 2010 and 2011.

The IP participants were eager to test the effect of the resulting fertilizers on their crop yields. When the time of harvesting came, we gathered in order to jointly evaluate the results. Everyone was convinced that the two most promising options were BM and LV, as these materials increased the availability of P the most and achieved highest crop yields. Another decisive factor was that BM and LV were locally available in ample quantities and at low or no costs for the farmers (Locher, 2011). As Manjit pointed out: "Through the participation in the rock phosphate trials, I encountered buttermilk as a simple and economical solution to increase the P supply to my crops."

Figure 8.2 Focus group discussion with farmers in 2009
Photo: Authors

Such promising first results motivated us to follow up with a second set of trials in 2012. A first study looked more closely at buttermilk and local vinegar. It tested different ratios of BM/LV to RP and experimented with different time periods of incubation. The study concluded that incubating RP with buttermilk in a ratio of BM:RP = 10:1 for a period of one week was optimal for increasing the efficiency of RP (Nyffenegger, 2012).

We also conducted a second study to look at different options available for improved farmyard manure (FYM) management. This revealed that the so-called "shaded shallow-pit system"[3] best conserved the quality of FYM. Furthermore, local farmers too preferred this system for the storage of their FYM (Gomez, 2012).

Fertilize an egg with a sperm and a baby will be born. In the same line we gave birth to the "rock phosphate-enriched-FYM" (RP-FYM) technology, by marrying the information of the two studies described above. We set up a demonstration shed for the production of RP-FYM (a high quality phospho-compost) at bioRe (Figure 8.3). It works as follows: incubate one part of RPh with ten parts of buttermilk for one week, and then spread the mixture on 40 parts of FYM (Figure 8.3). In order to reduce nutrient losses, keep the RP-FYM on a tarpaulin foil and use the foil to cover it. Shade the whole structure to protect it from the sun.

An unexpectedly rapid evolution

So far so good, we had developed a technology, but would it really lead to higher yields? Five farmers were particularly interested in both the technology and an answer to the latter question. That is why we made them our lead

Figure 8.3 Farmers being trained in RP-FYM production (top) and demonstration shed for training (bottom)

Photos: Authors

farmers. We gave them the first batch of RP-FYM we had produced at bioRe, and they used it to set up trials in their wheat crops in 2012–13. At the same time, we built sheds on their farms, and they started to produce the phospho-compost by themselves. At the end of the season, we discussed and evaluated the results with them using a farmer field school approach (Figure 8.4).

These five farmers were our ambassadors. We built five teams of five with one lead farmer and four associated farmers per team. The lead farmers acted as team leaders, teaching their associated farmers how to produce the new fertilizer (Figure 8.4), and showing them how to put up trials in their fields.

Figure 8.4 Exchange visit with five lead farmers to evaluate the effects of different fertilizer treatments on yields of wheat grown in 2012–13 (top) and RP-FYM production on a lead farmer's farm (bottom)

Photos: Authors

Each of the lead farmers produced enough RP-FYM to supply his associated farmers with batches for them to set up trials in cotton, soybeans and wheat on their own farms in 2013–14. This way, we managed to carry out a total of 37 on-farm trials.

The results of these trials outperformed the expectations of all the IP participants: the yields of cotton, soybeans and wheat all increased significantly in the RP-FYM treatment (in some cases by more than 100 percent) as compared to farmers' practice. On average farmers harvested some 40 percent more

Table 8.1 Yield increases (mean ± s.e.m.) in on-farm trials conducted in 2013–14

Crop	Farmers' practice (kg/ha)	RP-FYM treatment (kg/ha)	Increase (%)	Number of farmers/ trials (=n)
Seed cotton	1,170 ± 205	1,646 ± 222	41	10
Soybeans	1,548 ± 118	2,163 ± 227	40	14
Wheat	2,758 ± 219	3,138 ± 242	14	13
Mean	1,825 ± 151	2,316 ± 165	31	37

s.e.m.: standard error of the mean ($\sqrt{n}$).

White Gold and soybeans (Table 8.1). These results were consistent across different types of soils (high/medium yield potential soils) and farms (smaller/ bigger farms). We received reports that these effects are also consistent across years, as for instance Manjit told us that he continues to harvest around 33 percent more cotton with RP-FYM to date. But this success did not come about by chance. Between 2009 and 2014 we had set up a total of 159 RP-FYM trials with 118 farmers from 31 different villages. A man reaps what he sows.

The participating farmers were very pleased with the results they had achieved, and promptly engaged in more creative thinking, brainstorming how the technology could be further developed. bioRe India Ltd. also reacted positively to the results:

> The rock phosphate trials are one of the best examples we have from our participatory research activities. It improved the knowledge of both our extension teams and our farmers, while it also allowed for the conservation of traditional farmers' knowledge.
>
> (Mr. Vivek Rawal, CEO and director of bioRe India Ltd)

Does switching to organic pay off?

Besides the phospho-compost success story, the LTE also led to valuable results: as expected, organic cotton systems showed lower yields, by 10–15 percent. But the production costs were also lower, by 40–65 percent (Forster *et al.*, 2013). So at the end of the day, the organic and the conventional farmer have the same amount of money in their pockets. Why does organic pay off then? Because the organic farmer took less risk; he invested less money to grow his crop which can have important implications in cases of crop failure.

Sitaram Thakur, president of bioRe Association, stressed the importance of this information: "The involvement of farmers in the LTE helped to clarify many open questions about organic farming, and provided us with an opportunity to make an unbiased choice about the type of production system we wanted to engage in." And the farmer Rajendra Singh Mandloi underlined:

Table 8.2 Results of validation trials conducted from 2009 to 2013

Year	Number of trials[a]	Number of farmers with trial(s)[b]	Number of farmers involved[c]	Number of farmers who joined bioRe[d]	Percentage of involved farmers who joined bioRe	Number of farmers who joined bioRe per farmer with a trial
2009	18	10	45	0	0	0.00
2010	30	21	90	0	0	0.00
2011	59	49	178	55	29	1.12
2012	53	53	208	150	72	2.83
2013	55	55	210	53	25	0.96
Average 2011–2013	56	52	203	86	42	1.64

[a] Trials were carried out in cotton, soybeans, wheat and chickpeas.
[b] Number of farmers with trial(s) may be lower than Number of trials due to several trials of a single farmer.
[c] Number of farmers involved includes farmers with trial(s) and visiting farmers (exposure visits).
[d] Number of farmers who joined bioRe includes farmers with trial(s) and visiting farmers (exposure visits).

> Before the existence of this innovation platform, there was a lack of information. I was doing organic farming on my own, and I was desperately looking for any authentic source of information. This platform has filled this gap and served as a milestone for organic farmers in the region.

The LTE and bioRe concepts also attracted the attention of conventional farmers. They wanted to see the performance of organic cotton on their own farms. "OK" we thought, and, taking advantage of our IP, launched another subproject: the validation trials. Did the LTE findings reflect the real situation of farmers on the ground? Yes they did. During the first two years (2009 and 2010), conventional farmers were not convinced, because commonly observed yield depression during the conversion period to organic farming (Panneerselvam *et al.*, 2012) also became manifest on their fields. However, they started recognizing the benefits of organic farming from 2011 onwards, and many of them joined bioRe: per farmer with a trial, an average of 1.64 farmers joined bioRe from 2011 to 2013. In 2012, the number reached almost three farmers per farmer with a trial (Table 8.2), which clearly underlines the potential of validation trials and exposure visits with neighboring farmers.

Creating impact through IPs

Competitions to stimulate excellent performances

How to increase the yields of 5,000 farmers by 30 percent? Good question, especially because that's the stipulated target impact of our IP. We needed knowledge transfer. Knowing that the building materials for each RP-FYM shed cost about 100 USD, we quickly realized that building many more sheds would have been too expensive. We were in desperate need of a smart idea in order to reach the farmers who had not been involved in our activities.

Many great men made it into the books of history because they had a deeper understanding of only two words: spontaneity and intuition. They listened to their guts, and this is rarely the wrong thing to do. The director of FiBL advises his employees to take their coffee breaks, as they are at the root of most innovations. In our case, the flash of inspiration struck in a meeting at bioRe: an IP member came up with the brilliant idea to launch a competition among the participating farmers. We asked them to initiate the production of their own RP-FYM; whether it was in a shed similar to the ones we had built or underneath a tree and covered with palm leaves didn't matter.

Just like the CGIAR Research Program on Integrated Systems for the Humid Tropics launched its IPs Case Study Competition which lies at the root of the text you are just reading, we announced a valuable award in order to stimulate creativity and superb outputs of our farmers: a cow with its calf for the most innovative idea or the best quality phospho-compost. The word about the competition spread like wildfire, reaching many more farmers than the project could have ever informed. Rajesh Shobharam, for instance, built a low-cost shed from scrap materials he found lying around his yard.

The air vibrated with excitement during the period leading to the award. Many farmers must have had thought "is my idea good enough to beat my neighbor?" during these weeks and months. In the meantime, the project team was busy preparing for the award ceremony: we prepared illustrated leaflets in English and Hindi, printed posters and drafted a laudation for the winner. In order to avoid controversy, we needed to make a fair judgment based on objective assessment criteria. To identify the winner, we decided to rate the nutrient contents analysis of the RP-FYM the participants had produced. The farmers agreed with this procedure, so nothing stood in the way for an enjoyable award ceremony.

Finally, the day of the ceremony came. Farmers screened their wardrobes for the nicest set of clothes, everyone was excited and the event attracted considerable attention. In total, 96 farmers and 12 bioRe and FiBL staff participated. Rajesh Shobharam was announced as the winner, and he humbly accepted his prize (Figure 8.5). After the laudation, we gave the floor to him:

> When I collected the manure, I was not doing it with the intention to win the competition. But shortly before the ceremony I felt I had done a good job, as I had strictly followed the instructions my lead farmer Manjit gave me. My manure was very good, and I achieved a high crop yield, so I had a good feeling.

We also gave consolation prizes to all the other participants in order to acknowledge their commitment and good results. They sincerely thanked us

Figure 8.5 Competition award ceremony with participating farmers, researchers and extension agents. The best quality phospho-compost (RP-FYM) won a cow and calf. Competition winner Rajesh Shobharam (front middle with flower neck garland) and project leader Dr Gurbir Singh Bhullar (front middle, wearing turban)

Photo: Authors

for organizing "one of the best activities the project has ever carried out," and encouraged us to launch more such competitions.

Going viral: the power of simplicity

Was this ceremony not the perfect opportunity for further dissemination? Yes it was. We just had to take advantage of having so many ambassadors in one place at the same time. Together, they could cover all the 5,000 fellow farmers of Manjit. Especially the extension agents responsible for each extension center in the area surrounding the IP had the potential to make our interventions go viral. We provided them with leaflets and posters, and encouraged everyone to further spread the information by word of mouth. After a proper feast they departed, eager to go back to their districts in order to build demonstration sheds and train farmers. We have received oral reports that these facilities have been used for training farmers and sharing knowledge and experiences ever since. However, at the end of the day the most powerful tool that led to the adoption of the technology was also the simplest one: farmer to farmer extension. In other words: word of mouth.

Why was this award such a success, such an impactful event? Because we did not have to start from scratch; flashback: we had performed 39 meetings,

27 exchange visits on participating farmers' fields and 27 workshops around this topic between 2009 and 2014. And of course, we had our LTE next to which we also installed RP-FYM trials for exposure visit; 437 men and 339 women participated in these events in 2013 and 2014 alone, figures that emphasize the potential of exposure visits with neighboring farmers. We repeat: "a man reaps what he sows."

The future of organic White Gold

We don't rest on our laurels. Just like a company who launches a product, you constantly have to adapt in order to keep up with the changes around you. There is potential for improvement of the RP-FYM technology. One farmer, for instance, came up with the suggestion to simultaneously mix in wood ash in order to enhance the potassium content. Moreover, the socio-economic sustainability has to be further investigated: how much buttermilk is available for the farmers, and at what time periods? What if there is a market to sell the buttermilk? How about the availability of RP in the villages, and the sustainability of RP in general? After all, RP from phosphate mines is a finite resource that cannot be manufactured (Neset and Cordell, 2012). And last but not least, what do the market characteristics and dynamics of both FYM and buttermilk look like?

What's next for our IP? Besides the fact that we need to assess the impact of our interventions on the livelihoods of our farmers (after all, raising agricultural productivity is just one of the five pillars to improve the income and food security of poor people in low-income countries (GAFSP, 2014), we are going to address the big challenges for the production of organic White Gold; Organic pest control, for example, is still one of the major constraints. But beyond that, arguably the most daunting issue is the lack of suitable seeds. As breeding companies focus almost exclusively on *Bt* cotton hybrids, organic producers are increasingly cut off from the progress in breeding. We had to do something about that, so we used the IP as a stepping stone to launch yet another project about breeding: since 2013, "Green Cotton"[4] has been pursuing the objective of training farmers on how they can sustainably cover their seed demand by themselves. Will we be able to contribute to sustaining the supply of organic White Gold from India? We strongly believe so.

It is not about best practice, but best fit

This case illustrates the advantage of combining applied science with participatory action research. Agricultural systems are complex and unpredictable. Accordingly, we cannot hope to simplify the development processes of such agricultural systems. Instead, as our case demonstrates, we can harness this complexity to our advantage. How? You teach a number of people the underlying principles of your innovation, and ask them to implement it by themselves. You'll be surprised by the many different ways they take and

WHITE GOLD (INDIA)

all these different ways can together lead to a better end result. Our point of reference was a long-term farming systems comparison trial (LTE), an agronomic on-station experiment. As we did not have any limitations in financial or human resources in this trial, we were able to ensure optimal management conditions for the crops we grew. The resulting crop yields were higher than the average yield of the farmers we worked with in our IP. Since the productivity increase due to our innovation in the LTE was consistent with the results our ambassador farmers had obtained in their own field trials, we did not have to try hard to "sell them" on our innovation to other farmers. They readily embraced and adopted the technology on their own.

The diversity of our approach made the farmers confident: they could test out new technologies on their farms and exchange their experiences with us at the central LTE, as well as on their farms. As the bioRe extension agent Randhir Chohan pointed out: "The combination of participatory research and long-term experiment provided a scientific basis which helped us to provide authentic knowledge to our farmers." The greatest lesson of our experiment was: if we teach farmers to carry out research on their own farms, they are more eager to own their innovation, adopt it in practice and spread the innovation by word of mouth. All these processes can eventually lead to a snowball effect and thus considerable impact.

How can we bring the successes described in this case study to scale and help those who want to start a similar IP for another crop? Our best practices of including farmers in research and allowing divergent methods of experimentation definitely are a good starting point. Of course, we recognize that each new replication of our model must be customized to local conditions. Yet some of the general principles we have touched upon in this case stand no matter what the local context. For instance, it is only when you have the general picture (such as results from meta-analyses), that you can break it down to the local level again.

Thus, it is our suggestion that International Agricultural Research for Development (IAR4D) needs to reinvent itself. If we are to bring our interventions to scale in order to create impact, we need a paradigm shift: IAR4D has to become IAR-*IN*-D, that is, International Agricultural Research-*IN*-Development. What is IAR-IN-D? It is a process of embedding scientific research in economic development by shortening the feedback loops that are inherent parts of innovation cycles, and involving farmers in real-time research and impact analysis. We need to honour the complexity of the systems we are dealing with through the research design of our projects. Social and natural sciences need to be integrated not only in our activities, but also in new forms of educational institutions. This direction is not only appealing to donors, but also to farmers, who can, at last, discover, enjoy and benefit from the process of IAR4D.

Acknowledgments

The field and desktop work of the whole bioRe Association team is gratefully acknowledged. In particular we would like to extend our thanks to Vivek Rawal, Sitaram Thakur, Ishwar Patidar, Akilesh Patak, Yogendra Shrivas, Rajeev Verma, Bhupendra Sisodiya, Lokendra Chouhan, Dharmendra Patel, Rajeev Baruah and Ritu Baruah. We are grateful to all our POR farmers and extension workers for their continuous commitment and good results. We thank Manjit Singh Dang, Rajendra Singh Mandloi, Rajesh Shobharam and Randhir Chohan in particular. This case study would not have been possible without the hard work of our colleagues at FiBL: thank you Christine Zundel, Dionys Forster, Monika Messmer and Paul Mäder. Last but not least, our sincere acknowledgement goes to our donors Biovision Foundation for Ecological Development, Coop Sustainability Fund, Liechtenstein Development Service (LED) and the Swiss Agency for Development and Cooperation (SDC) for their continuous financial support and commitment to long-term research.

Appendices

Appendix 8.1 Precipitation of phosphorous ions under alkaline soil conditions

The form in which free P ions in the soil solution occur depends on the pH. At pH levels below X you find PO_4H_3, at pH levels between X and X you find $PO_3H_2^-$, at pH levels between X and X you find PO_3H^{2-} and at pH levels above X you find PO_3^{3-}. Under the soil conditions described in this case study, you mostly find PO_3H^{2-}. These ions can be bound to free Ca^{2+} ions which are also found in the soil solution under alkaline conditions. If this happens, $CaPO_3H$ is precipitated, a process called "precipitation" (Hopkins and Ellsworth, 2005; Dick, 2007). For further information on P dynamics in the soils consult Marschner (2012).

Appendix 8.2 Details about case study stakeholders

In 1991, the Swiss yarn trader Remei AG and the Indian spinning mill Maikaal Fibres (India) Ltd. initiated the Maikaal bioRe® organic cotton project. What had started as a non-commercial experiment to help cotton producers find a way out of debt and secure a sustainable livelihood has meanwhile developed into an enterprise that joins social responsibility and ecology with economic profit. Maikaal bioRe, these days known as bioRe® India Ltd., has grown to become one of the largest and most well-known organic cotton projects worldwide, with more than 5,000 smallholders (figures year 2012–13) producing organic cotton and other organic commodities. bioRe distributes the needed inputs (e.g. seeds, organic fertilizers, pesticides, biodynamic preparations, etc.) to its farmers and purchases their cotton which is subsequently processed in bioRe's own modern ginnery.

Besides the commercial body of bioRe, the non-profit organization bioRe Association is an NGO that runs several social projects. These include a center for training and education that provides extension to local farmers and carries out research. The association also provides credit to farmers in order to promote infrastructure development (e.g. drip irrigation, biogas facilities, etc.).

Manjit Singh Dang represents the small-scale farmers associated with bioRe. bioRe assures market access for its farmers by a five-year purchase guarantee with a premium price of 15 percent for organic quality. In addition, Manjit and his fellow farmers regularly receive training in organic and biodynamic farming and participate in the ongoing research activities of bioRe Association.

In its early days, bioRe did not engage much in research due to the non-commercial nature of the initial project, and the subsequent direction towards sourcing of organic cotton aiming at the buildup of a steady supply chain. However, the need to engage in research became increasingly evident when the impacts of *Bt* cotton introduction started to become manifest. Subsequently, a close collaboration between bioRe and the Research Institute of Organic Agriculture (FiBL) was established. FiBL is the world's leading research institute on organic agriculture.

Following the chain of the White Gold, the Swiss yarn trader Remei AG exports and processes it into trendy clothing and other cotton products, many of which are sold by upmarket brands including "Naturaline" of Coop, the biggest retailer of organic products in Switzerland.

Notes

1   www.systems-comparison.fibl.org/
2   www.bioreassociation.org
3   FYM is stored in a shallow pit whose interior is covered with a thick foil. The pit is covered with a polythene sheet and the ground is slightly sloped in order to collect the effluent.
4   Funded by the Mercator Foundation Switzerland: www.greencotton.org/?lang=en

References

Bachmann, F., 2012. Potential and limitations of organic and fair trade cotton for improving livelihoods of smallholders: evidence from Central Asia. *Renewable Agriculture and Food Systems* 27, 138–147.

Choudhary, B.G.K., 2010. Bt cotton in India: a country profile. ISAAA Series of Biotech Crop Profiles. ISAAA Ithaca, New York.

Dick, W., 2007. Biochemistry of phosphorus and sulfur transformations in soil. Available online: http://senr.osu.edu/sites/senr/files/imce/files/course_materials/enr6610/Section 06_Text.pdf (accessed 5 February 2015).

Forster, D., Andres, C., Verma, R., Zundel, C., Messmer, M.M., Maeder, P., 2013. Yield and economic performance of organic and conventional cotton-based farming systems: results from a field trial in India. *PLoS ONE* 8, e81039.

GAFSP, 2014. Global Agriculture and Food Security Program, Project Implementation Update. Available online: www.gafspfund.org/sites/gafspfund.org/files/Documents/ImplementationUpdate_Feb%206_FINAL%202.pdf (accessed 20 March 2015).

Gomez, S., 2012. Identification and evaluation of improved manure management options in the context of rural india. BSc Thesis. Swiss College of Agriculture, Zollikofen, Switzerland.

Hopkins, B., Ellsworth, J., 2005. Phosphorus availability with alkaline-calcareous soil. Available online: http://isnap.oregonstate.edu/WERA_103/2005_Proccedings/Hopkins%20Phosphorus%20pg88.pdf (accessed 5 February 2015).

IAASTD, 2009. International assessment of agricultural knowledge, science and technology for development (IAASTD): Executive summary of the synthesis report. Available online: www.unep.org/dewa/agassessment/reports/IAASTD/EN/Agriculture%20at%20a%20Crossroads_Synthesis%20Report%20(English).pdf (accessed 20 March 2015).

Kanwar, J., 1988. Farming systems in swell-shrink soils under rainfed conditions in soils of semi-arid tropics. In: Hirekerur, L.R., Pal, D.K., Sehgal, J.L., Deshpande, C.S.B. (eds), *Transactions of International Workshop on Swell-Shrink Soils*. National Bureau of Soil Survey and Land Use Planning, Nagpur, India, 179–193.

Locher, M., 2011. Options for rock phosphate solubilization in organic farming and their effects on mung, wheat and maize. BSc thesis. Swiss College of Agriculture, Zollikofen, Switzerland.

Mäder, P., Fliessbach, A., Dubois, D., Gunst, L., Fried, P., Niggly, U., 2002. Soil fertility and biodiversity in organic farming. *Science* 296, 1694–1697.

Marschner, H., 2012. *Marschner's Mineral Nutrition of Higher Plants* (3rd edn). Academic Press, San Diego, CA.

Neset, T.-S.S., Cordell, D., 2012. Global phosphorus scarcity: identifying synergies for a sustainable future. *Journal of the Science of Food and Agriculture* 92, 2–6.

Nyffenegger, M.R., 2012. Improving plant availability of p contained in local rock phosphate for use on alkaline soils. BSc Thesis. Swiss College of Agriculture, Zollikofen, Switzerland.

Panneerselvam, P., Halberg, N., Vaarst, M., Hermansen, J.E., 2012. Indian farmers' experience with and perceptions of organic farming. *Renewable Agriculture and Food Systems* 27, 157–169.

Truscott, L., Denes, H., Nagarajan, P., Tovignan, S., Lizarraga, A., Dos Santos, A., 2013. Farm & Fiber Report 2011–12. Available online: http://farmhub.textileexchange.org/upload/library/Farm%20and%20fiber%20report/Farm_Fiber%20Report%202011–12-Small.pdf (accessed 4 February 2015).

9 MilkIT innovation platform

Changing women's lives – one cow
and one litre of milk at a time –
deep in the foothills of India's
Himalayan mountains

*Thanammal Ravichandran, Nils Teufel and
Alan Duncan*

We are in need of such platforms to find the target communities to get the impact very fast
(T.K. Hazarika, General Manager, Uttarakhand, National Bank
for Agricultural and Rural Development NABARD)

Introduction

In 2012, Tulsi Devi, a 39-year-old widow from the Baseri village in the Himalayan hills of Uttarakhand, India was left struggling to make ends meet. Her husband had died a few years back after a prolonged battle with alcohol addiction. She found herself with just one indigenous cow and a buffalo and a small piece of land barely large enough to produce sufficient rice and wheat to feed her family. The distance from her village to the nearest mountain road leading to the local market made it impossible to sell her surplus milk. She struggled even to pay school fees for her children. Seeing no other option, she sent her eldest son, Sunder, who was only 15, to Delhi to work in a factory.

Tulsi Devi's life became easier when she joined the MilkIT innovation platform (IP) meeting in January 2013 that created an opportunity to interact with stakeholders to find new ways for selling milk. The regular income flowing in her home gave her the confidence to send her remaining children to school.

The IPs formed in the beginning of 2013 by ILRI helped to address the issues of 1,244 families similar to those faced by Tulsi Devi. The efforts made by the platform set in motion a series of events that led to stronger milk sales, rapid adoption of feed improvement practices and increased milk production. A review of the IPs at the end of 2014 showed that the platforms have facilitated increased incomes for more than 600 households, improved collaboration among the local development institutions, provided employment for many women and that the platforms have changed the mindset of various development policy makers. Mr Ahmed Iqbal, the Chief Development Officer of Almora district has said:

MilkIT platform caught me at the right moment. It seemed to be a catalyst to do something; it also showed that small interventions really can make a difference. So we could really scale this up. I found something that really needs a trigger to have wider results.

Women in remote hill villages struggle to develop dairy as source of income

The State of Uttarakhand is characterized by subsistence-oriented mixed agriculture with dairy farming. However, opportunities for generating income are limited, resulting in considerable out-migration to nearby cities especially among men. Women play an important role in dairy farming, but most milk is consumed within households themselves or given to relatives free of cost. Women walk long distances every day among the steep forest hills to collect fodder for their cows and buffaloes. Despite their efforts, these women are not receiving any cash income from their dairy animals. However, improved infrastructure, in particular road connectivity, has in recent years created opportunities for these farmers to link to larger markets and has thus increased the potential to generate income from dairy farming. Nevertheless, farmers still face high transaction costs due to low milk production and the considerable distance of some villages to paved roads. Improved feed and breeding technologies promoted by various institutions have not been widely adopted, as they have generally not been tailored to women's requirements and have not considered market linkages.

Grounding the IPs

Pull villages together for collective action

Although villages and settlements are typical units for identifying development activity areas, we decided that IPs would require larger units to trigger collective actions, a decision that was supported by our experience in the district of Bageshwar. Focusing on larger geographical units attracts non-producer stakeholders such as the private sector, especially where dairy value-chain development is concerned. Therefore, village clusters were formed to serve as activity units for IPs. The project was implemented in two districts, Almora and Bageshwar. In each district, two IPs for feed innovations, covering 4–6 villages each, were combined into one market IP for strengthening the market linkages (Figure 9.1). The IPs covered 1,244 families in 21 villages (Table 9.1).

Seeking members for the IP: intervention history exercise

While setting up the IP, finding the right institutional members is important. Detailed interviews with key personnel of government and private development

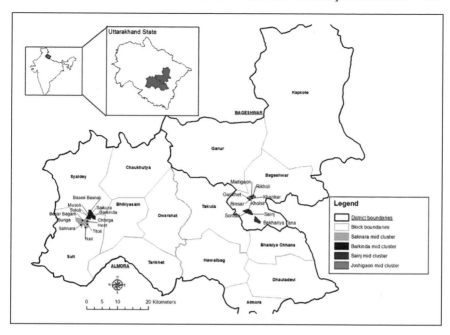

Figure 9.1 MilkIT project IP clusters in Almora and Bageshwar districts, Uttarakhand State, India

Source: Subedi *et al.* (2014)

Table 9.1 Feed and market IP cluster composition

District	Name of market IP	Name of feed IP	Number of villages	Number of families
Bageshwar	Bageshwar	Saing	4	379
		Joshigaon	6	243
Almora	Sult	Saknara	6	379
		Barkinda	5	243
		Total	21	1,244

Source: Own research

organisations and NGOs focusing on the intervention history of the last decade provided the necessary insights to select appropriate institutional members for the IPs. The selected development actors included the state dairy cooperative (Aanchal, representatives of district and block level), staff of the IFAD-supported development programme, financial institutions (commercial banks, development bank), BAIF (national NGO for breed improvement), development NGOs, district animal husbandry department and extension services (Krishi Vikas Kendra (KVK)). We had expected the agriculture and forest departments to be keenly involved, but we found that they were not able to participate regularly in IP meetings. When asked later, they explained that the development of the dairy sector was not among their key priorities.

How can the IP reach more farmers?

There are various options in the formalisation and modes of communication while organising IP meetings (Nederlof *et al.*, 2011). Within MilkIT, the aim was to reach large numbers of farmers and stakeholders through the IP approach. Three types of meetings were organized.

First, core IP meetings were organized for each IP on feed and market issues in which representatives of producers and non-producers participated, despite the difficulties of including all stakeholders (Steins and Edwards, 1999).

Second therefore, there was demand for follow-up meetings at village or cluster level to address the issues including the following:

* Only a few representative farmers were able to participate in the IP meetings and there was need for dissemination of discussion/information at village level to allow many farmers to take collective decisions.
* Actions that had been agreed in IP meetings required follow-up at village/settlement level and with individual institutions.
* Some issues differed between villages and therefore needed further discussion at village level.
* Village level meetings provided more opportunity for farmers to express their views which were then taken back to core IP meetings.

The third type of meetings consisted of exchange visits and participatory training sessions that helped with building the capacity of farmers in applying improved technologies and practices and in many cases initiated the adoption of proposed innovations.

Table 9.2 Summary of MilkIT IP meetings (December 2012–July 2014)

Type of IP meeting	Sult (no. of meetings)	Bageshwar (no. of meetings)
Market (IP core)	4	3
Feed (IP core)	2	2
Follow up in villages (market and feed)	53	149
Training/exchange	1	3
Individual institutional	2	5
Total	62	162

Source: Own research

All the meeting discussions and follow-up actions were recorded and stored on a shared platform (Google Drive). A summary of meeting numbers over the 20-month project period is given in Table 9.2.

Constraints, achievements and overcoming the challenges

Identification and prioritisation of the common issues or constraints for the selected development topic (e.g. dairy development) is an important first step to enable the effective functioning of any IP (Nederlof *et al.*, 2011), and generally requires effective facilitation. This project followed several participatory approaches to prioritise major issues. Before IP formation, focus group discussions with producers and non-producers using the FEAST tool (feed assessment tool developed by ILRI) and semi-structured interviews with key stakeholders helped to understand the issues from the producer perspective. These issues were discussed again in the initial IP meetings using participatory discussion methods to prioritise the key issues. It was an interesting experience to see different innovations emerging from these discussions to address similar issues depending on local context.

Constraint 1: Small villages, long distance, where to sell little milk?

Since distances to the next road were long and only a few dairy animals were kept, each producing low milk yields, the transaction costs for milk marketing were prohibitive. The only option to sell milk for these farmers was through the state dairy cooperative 'Aanchal' which was subsidising transport by paying for people to transport the milk from the village to the paved road on foot. However, the cooperative's reach to remote villages was limited, covering only a few villages. No efforts were taken to expand this arrangement to other villages interested in selling their milk. Several dairy collection centres had been started by the state cooperative but were closed after a few years. No effort was made to identify the reasons for the failure of dairy collection centres in these villages. A small study initiated by ILRI in 2013 found that farmers had stopped selling milk to these institutions for a range of reasons including, among others, uncompetitive milk prices, inappropriate targeting of beneficiaries (credit support), rigid rules requiring a minimum of 30 members from each village, and governance issues in measuring quality of milk.

Solution 1: 'Let us come together to sell milk and strengthen the system'

The first intervention adopted by the participants after the initial IP meetings was to improve market links. In Bageshwar, farmers requested an improved price and monitoring system from the state dairy cooperative, Aanchal, as they felt the price they were receiving for their milk was too low. Aanchal failed to address this issue by the time of the next meeting, which led the farmers to set up Jeganath dairy cooperative, an independent self-help group cooperative covering 8–10 villages. As many farmers in these villages had already been organized into self-help groups, it was easy to bring them together for this initiative. Initially, only 32 farmers participated in the Jeganath dairy cooperative in April 2013, but this soon increased to reach more than 100 farmers in six months. The farmers established a shop in Bageshwar, the nearby town, and contracted a private vehicle for collecting milk from villages. In each village, individuals, such as Geeta Bisht in Kolseer village (Figure 9.2), were elected as group secretaries to collect milk, receiving INR 2/litre as their incentive. The IP members fixed an appropriate milk price based on milk quality.

On the other hand, farmers from Sult preferred to improve their links with Aanchal, the state cooperative, as the distance from their settlements to any town is far. Four new collection centres were formed in this block collecting milk from eight villages. Before the IP, Aanchal had insisted on a minimum of 30 signed-up households in each settlement for establishing one collection centre. However, many settlements in this area consist of less than 20 households. This issue was discussed in IP meetings and Aanchal directly and as a result, Aanchal relaxed this rule and is now allowing 2–3 settlements to form one village cooperative together.

Figure 9.2 Geeta Bisht is now employed in Kolseer village, Bageshwar, to collect milk
Photo: ILRI/T. Ravichandran

Solution 2: We can help to increase production-motivated actors

Identifying effective solutions and implementing agreed actions depends to a large extent on the motivation of the involved actors. Generally, each actor will have their own specific motivation to participate in IPs. It is a major task of IP facilitation to elicit and match these motivations. In this case of MilkIT this was most obvious in regard to credit issues. Once the improved marketing arrangements for milk had been established, many farmers, especially men supported by their women, expressed their interest in purchasing high-yielding dairy animals. However, due to the multitude of formal requirements they could not receive any credit from their regular banks. The IP members from the finance sector, private banks and NABARD, the national bank for agriculture and rural development, came forward to address this issue as they could see a good opportunity to employ development-oriented credit facilities. A private bank appointed one coordinator at block level to reduce formalities. Furthermore, the option of group liability rather than asset liability was introduced, a considerable help for farmers with very little land or other assets. NABARD has subsidized the interest on loans to farmers who have been servicing their loans regularly for 12 months.

Solution 3: Overcoming power dynamics and taboos – how the MilkIT IP succeeded

Handling distorted power dynamics was a considerable challenge for the facilitator in the initial stages of the project. Where these dynamics are not addressed, they can seriously obstruct innovation processes (Cullen *et al.*, 2013). Farmers, especially women, were reluctant to express their views when IP meetings were conducted at government venues. The dominance of higher officials from various government departments led to 'preaching to farmers' rather than listening to their needs. Temples or community halls, which were subsequently chosen as meeting venues, offered women and small farmers a 'safe space' to voice their opinions. Farmers were then also able to invite development stakeholders to their nearby villages or houses to demonstrate actual practices. This allowed non-producer stakeholders especially from government bodies to gain a better understanding of the issues discussed and actions agreed, compared to merely attending meetings (Blackmore *et al.*, 2007).

Improving links to markets was the first and most important action taken by farmers, yet in a few villages farmers were very reluctant to sell any milk at all due to social and religious taboos. Some of them reported that 'selling milk is sin' or 'if I sell milk, others don't respect me'. We found these views such serious barriers for emerging innovation that our facilitators decided to stay in these villages for a few days. Their efforts paid off through identifying 'change agents'. For instance, when the facilitators reached out to and convinced Bhandari, a respected teacher in Besarbagarh village, he in turn persuaded many women to sell their milk. Now that the village receives an additional income of around USD700 per month, the teacher says the changes in their village towards a better life have become visible to all.

Solution 4: How to attract the private sector?

Stakeholder participation or membership is not fixed in the IP. At any given time new members can join the IP meetings depending on the needs identified as well as the opportunities and incentives created by the platform. Both dairy market platforms were finding it difficult to get private milk traders to participate in meetings and extend their milk collection. The trader's opinion was that 'these villages comprising 20 to 100 animals will not give us any profit because of the small volume of milk. We will be interested if there is more milk.' Since the support by finance institutions for purchasing cross-bred cows resulted in increased production, a private trader is collecting milk from Saing village in Bageshwar district where more than 100 litres are produced daily. These farmers are selling their milk partly to the Jeganath cooperative and partly to a private trader. These farmers' groups have also negotiated with a private feed company to receive concentrated feed at wholesale prices.

Challenge 2: How to manage the fodder scarcity?

Animal feeding in the Himalayan hills is dependent on grass collected from the forest area, which contributes about 70 per cent to livestock feed, with crop residues and tree leaves making up the remainder. Fodder collection and feeding are predominantly women's work in this area. Women collect fodder from forests, remote unused lands and the bunds of cultivated land. On average, this takes 3–4 hours per day. At the end of the rainy season women cut the forest grass for hay-making and store this for the lean periods in winter and summer. Despite all the efforts involved in fodder collection, a lot of fodder goes to waste due to feeding on the ground. Women estimated that 20–25 per cent of the fodder is wasted because it is stamped on by animals or gets mixed with urine and dung. When this was discussed in the IP meeting, it was found that a lack of knowledge on alternative feeding practices and a lack of financial resources were the main hurdles to improving this situation.

The second issue was the seasonal shortage of green fodder. The greater variability of rainfall during the last few years has resulted in increased scarcity of green fodder during winter and summer periods.

Solution 1: Participatory action research – how to reduce wastage

As a starting point we interviewed a few key farmers and development actors to better understand previous interventions. It was painful to see manually operated wheeled choppers distributed by several institutions rusting away unused. Shanti Devi from Garikhet village said that 'It needs two persons to

Figure 9.3 Shanti Devi, Garikhet village with a women-friendly chopper (simple knife and frame)

Photo: ILRI/T. Ravichandran

operate, I am the only one at home, and how can I operate this?' Based on discussions at an IP meeting, a low-cost, simple wooden handle knife and mechanical sickle choppers (Figure 9.3) were identified as appropriate and attractive implements for chopping fodder. A local manufacturer agreed to produce these choppers at a reasonable price. Finally, a cost-effective and simple feeding trough was designed according to the size of local animals with the help of partner staff.

Imposed technologies can hamper joint learning, whereas learning is a prerequisite for successful innovation (Kristjanson *et al.*, 2009). Initially, farmers were not convinced that the fodder savings through the feeding trough and choppers would outweigh the additional costs. However, participatory trials showed that the use of these improved technologies roughly halved fodder wastage and thereby provided 11 per cent more feed. The immediate benefit of saving fodder was reducing the burden to women; 90 per cent of farmers participating in these trials were women. They reported that these technologies reduced the time required not only for fodder collection from forests, but also for cleaning the waste around the animals. These results were shared in the feed IP meetings that initiated the adoption of these technologies on a wider scale by many farmers. Participating stakeholders including IFAD and NABARD helped with a subsidy (50 per cent) for constructing the feeding troughs and for purchasing the choppers. This helped with the uptake of these innovations. More than 130 farmers constructed feed troughs and more than 225 farmers adopted the women-friendly choppers in one year.

Solution 2: Increase fodder production – dual purpose cereal crops, improved forages

To increase the availability of green forages during the lean periods of winter and summer, technical partners including the local extension service (KVK) and ILRI suggested in the IP meeting the introduction of dual purpose crops (food and feed), temperate grasses and improved forages such as Napier and clover. Demonstration plots for dual purpose crops such as wheat, barley (allowing an early cut during the vegetative stage without affecting grain yields) and maize (providing large amounts of nutritious stover) led to a wide adoption by farmers on small land parcels. Napier grass is promoted by many organisations but its adoption is limited to areas with considerable rainfall or other water resources.

Challenge not addressed: where to get seeds?

The main problem in introducing improved grasses was the sourcing of seeds that were not available from the participating stakeholder institutions. In addition, identifying appropriate grass species was challenging due to extreme weather conditions, including cold winters, dry and hot summers and tropical rainy seasons. There is very limited institutional support for grassland improvement by state institutions. Establishing village-level seed multiplication systems was beyond the scope of the platforms during the short project period. Although farmers were happy with the additional fodder produced with dual-purpose cereals (wheat, oat and barley) the price of seed supplied by the KVK (50 per cent higher than regular cereal seed) may limit the sustainability of this intervention.

Efforts and actions of IP led to impacts

Increased income and employment

'Small initiatives can make a big difference': The Jeganath dairy cooperative created by the Bageshwar IP has had a strong impact on the livelihoods of many individuals. Along with Geeta Bisht (pictured in Figure 9.2), seven other people including four women are employed in milk collection, transport and retail. However, the greater effect of the cooperative is probably the opportunity for over one hundred farmers to earn INR 600 to 6000/month through milk sales. Most of this income is handled by women who use it to pay for household expenses, school fees and the purchase of feeds. In Sult region, more than 100 women like Tulsi Devi and their households are benefiting from the dairy collection centres established by the state cooperative. Devki Devi from Besarbagarh village said that 'Now I earn more than 1500 rupees per month through transport of milk from my village to the road. This income is helping me to get nutritious food for my kids and builds my confidence'.

A preliminary impact study conducted in November 2014 has provided evidence that families participating in IP meetings have five times more savings through milk sales than non-participating households. Over a 12-month period, farmers participating in IP meetings have fed their animals with improved forage for 50 days whereas non-participating households have only had forage for 12 days.

Increased communication

Improving communication is a core aspect of IPs in general and was one of the major components of the MilkIT project. In reviewing the project's success in this regard several aspects stand out.

When initiating the IPs it was apparent that smallholder producers already had a strong tradition of group formation and within-village communication, including a strong voice for women. This greatly helped with identifying producer representatives and with the feedback of IP meeting results back into villages. However, these groups, especially the women among them, regularly reported that never before had they had the opportunity to communicate with representatives from other villages and with higher-level representatives of stakeholder institutions.

Stakeholder institutions also valued the opportunity to engage with larger groups of development-oriented smallholder producers through structured dialogue. They appeared to view IP meetings as an efficient access route to their target populations. They also appreciated the communication products generated by the project and integrated them into their activities. On the other hand, it was not clear how far stakeholder institutions valued the opportunity of increased communication among themselves. Greater coordination among development actors leading to greater efficiency and impact does not seem to feature strongly in stakeholder assessments of the IP approach. Rather, queries were raised even within the project whether IP meetings should only be seen as an initial stimulus for increased bilateral communication between producers, development organisations and market institutions, questioning the sustainability of the IP approach.

However, the greatest challenge in improving communication appeared at state level. It was a stated aim of the project to integrate the project into the larger development framework and this was attempted through the establishment of an advisory council. While the six-monthly meetings provided regular updates on the project's progress to state-level representatives this did not appear to lead to greater interaction of the participating institutions with the project. Most improvements in interaction seemed to be at district level. Perhaps district-level changes have to become apparent first, before state-level representatives begin to take serious interest.

Factors contributing to impact

In reviewing the changes stimulated by this project and the contributing factors, three levels of contribution appear to be important.

Box 9.1 Meet Mahesh Tiwari who doubled his income through Jeganath dairy cooperative

Mahesh Tiwari is 23 years old and from Bolna Naghar village, Bageshwar district. For two years he was working in a Delhi factory after leaving school. Although his village was not selected for this project he started participating in the Bageshwar IP meetings. He soon joined the Jeganath cooperative formed after the initial meetings. The new business opportunities led him to reconsider his plan to work in Delhi to support his family. Instead, he applied for a loan from Aanchal to purchase cross-bred cows. This was refused but NABARD, the national  development bank, agreed to provide a loan with subsidized interest. He purchased two cross-bred cows and built a cattle shed with technical support from the KVK, the national extension organisation. Currently, he has increased his herd through purchasing two more cows with savings from his milk sales over the past 14 months. He is currently earning INR 12,000–15,000 per month (USD200–220), twice his factory wages. He can be seen as an informal innovation champion (Klerkx *et al.*, 2010), stimulating other farmers to engage in the dairy business as a livelihood option after seeing his success.

First, the basic interest of smallholder producers in generating income through dairy production was a fundamental requirement for any change to happen. Although this aspect was not considered during the selection of clusters, fortunately three out of four selected clusters were eager to increase their milk sales. One cluster realized, after some involvement, that the social issues involved with increased milk sales would not justify potential income benefits. During the selection of a replacement cluster, emphasis was placed on current income sources and interest in income development through dairy production. Clusters that already received most of their income from non-agricultural sources, where labour was very scarce and dairy production was not seen as a promising development pathway, were not considered.

Second, a supportive institutional landscape was essential to achieve wider impact. Over the course of the project, the assessment of which institutions would be interested and able to take up technologies and approaches identified by the IPs and contribute complementary interventions and resources evolved considerably. Some institutions that would have seemed natural scale-up partners did not seem to be willing to leave their established procedures, while others that were not specifically targeted developed considerable initiative. This was especially true for financial institutions that appear to have a role in stimulating change at least as significant as governmental and non-governmental development organisations. On the other hand, the general awareness by the state government of the potential of dairy development provided the necessary support to Aanchal to reassess its approaches to developing milk collection in remote areas.

Finally, the introduction of complementary technologies, both inputs and services, by active stakeholder institutions amplified the changes directly initiated by the project. Most obviously, this applies to the introduction of cross-bred cows, either through purchase or artificial insemination (AI), which enables a huge step in productivity.

What will be the future: forward linkages?

Within the project, the IP approach, an efficient process to identify and implement development interventions, is seen as the more important aspect compared to individual technologies or institutional arrangements. Various activities were undertaken to create a greater awareness among stakeholders of the procedures followed and the outcomes experienced. These included a sensitisation workshop during which an original drama on IP implementation was presented and a policy dialogue meeting at the state level.

This convinced the Chief Development Officer of Almora district to initiate monthly stakeholder meetings at the district level to address dairy development

---

Box 9.2 Conversations heard when a group of women evaluated MilkIT interventions

'I have no time to attend meetings' . . .

'Ho, it's painful to collect fodder and most of my fodder is wasted by this animal' . . .

'Let's try simple choppers' . . .

'I can sell milk now; I am employed to carry milk to the road' . . .

'Now people hear our voice' . . .

MilkIT DAIRY AND FEEDS INNOVATION PLATFORM (INDIA)

MilkIT DAIRY AND FEEDS INNOVATION PLATFORM (INDIA)

issues. It will be of great interest to follow the evolution of these meetings, especially in regard to participation, issues covered and procedures followed.

This project promoted many technologies and institutional changes including the following:

- The animal husbandry (AH) department has adjusted its policy formulation to include support for construction of fodder troughs, grassland improvement and improved buffalo breeding.
- Various organisations such as the AH department and IFAD loan projects have expressed their interest in promoting the adapted fodder chopper and feed troughs.
- The potential of dual purpose crops has been widely acknowledged by stakeholder NGOs and the AH department.
- The adaptation of village cooperative regulations to the local situation is being considered for wider application by Aanchal, as is the improved targeting of potential supplier communities and the realisation that improved monitoring and transparency of payment systems is required to regain the trust of smallholder producers.

Conclusions and way forward

The development of market aspects of dairy value-chains and the improvement of dairy feeding through IPs appears to be an effective and efficient approach to quickly stimulate impressive changes. It was important learning that the actual changes differed considerably between the various platforms, both in regard to value-chain development and feeding, highlighting the importance of leaving the prioritisation of interventions to the platforms themselves. On the other hand, supporting interventions through consistent documentation helped with their wider acceptance. Institutional changes in milk marketing appeared to be a major incentive for farmers to invest in feed and breed improvement despite increased input costs. It was obvious that especially in regard to feeding, simple interventions resulting in near-immediate benefits (such as fodder troughs and concentrate feeding) were more attractive initially than more complex packages with longer time horizons such as grassland development. However, the longer term effects of the IPs are probably more, due to better communication and collaboration of the various stakeholders. Enabling farmers to have their voice heard will allow for more efficient development efforts. Finally, IP partners who have identified various aspects of the project as valuable for their own activities are changing their approaches and are investing their own resources into wider dissemination. This has created an out-scaling potential that had not been envisioned at the project's outset. It would be very interesting to continue with the observation of how the established IPs evolve and how project outputs and experiences spread through institutions and into new geographical areas. Discussions are ongoing with various partners on how this could be achieved.

Acknowledgements

Authors acknowledge the funding support for the MilkIT project by the International Fund for Agricultural Development (IFAD). The authors acknowledge the staffs of partner NGOs, The Institute of Himalayan Environmental Research and Education (INHERE), Almora and Central Himalayan Rural Action Group (CHIRAG), Bageshwar for their support in the implementation of the project.

References

Blackmore, C., Ison, R., Jiggins, J., 2007. Social learning: an alternative policy instrument for managing in the context of Europe's water. *Environmental Science and Policy* 10: 493–498.

Cullen, B., Tucker, J., Homann-Kee Tui, S., 2013. Power dynamics and representation in innovation platforms. Innovation Platforms Practice Brief 4. ILRI, Nairobi, Kenya.

Klerkx, L., Aarts, N., Leeuwis, C., 2010. Adaptive management in agricultural innovation systems: the interactions between innovation networks and their environment. *Agricultural Systems* 103: 390–400.

Kristjanson, P., Reid, R.S., Dickson, N., Clark, W.C., Romney, D., Puskur, R., 2009. Linking international agricultural research knowledge with action for sustainable development. *Proceedings of the National Academy of Sciences* 9: 5047–5052.

Nederlof, S., Wongtschowski, M., van der Lee, F. (eds), 2011. Putting heads together: agricultural innovation platforms in practice. *Development, Policy and Practice Bulletin 396.* KIT Publishers, Amsterdam, the Netherlands.

Steins, N.A., Edwards, V.M., 1999. Platforms for collective actions in multiple-use common-pool resources. *Agricultural and Human Values* 16: 241–255.

Subedi, S., Thanammal, R., Cadilhon, J.-J., Teufel, N., 2014. Enhancing dairy-based livelihoods in India: mid-term progress report of the MilkIT project. Livestock and Fish Brief 5. ILRI, Nairobi, Kenya.

10 Synthesis

Jean-Joseph Cadilhon, Marc Schut,
Michael Misiko and Iddo Dror

Introduction

The introductory chapter of this book suggested a framework identifying different components that are necessary for an IP to achieve desirable outcomes or impacts. This framework assumed that platform support functions related to facilitation, organization, documentation and research can foster purposeful interactions between multi-stakeholder processes and content matter (see Figure 10.1 reproduced from the Introduction). The hypothesis was that with these three components in place, IPs could achieve various outcomes and impact: balancing trade-offs within complex agricultural production, marketing and natural resources management systems, transposing the innovation process from one product to work on an array of multiple commodities, and reaching out to a large number of beneficiaries.

Based on the case studies, this chapter will draw a synthesis of how these components of IPs are interrelated. This broad picture will also illustrate how the three components of IPs lead to outcomes and impact, based on the cases from this book. Table 10.1 identifies the case studies that are particularly strong illustrations of some of the four IP components, as judged by the editors of this book.

The case studies featured in this book demonstrate strong achievements in setting up multi-stakeholder processes, elaborating robust content matter or establishing well-designed support functions. What is more, all the platforms have also reached some form of outcome or impact that can be expected from mature IPs. The following sections in this chapter will illustrate how process and content of the IPs featured in this compilation have led to delivering impact. However, Table 10.1 also shows that none of the IPs in this anthology can boast of having achieved all three of the components necessary for success. Likewise, none of them has yet attained all three outcomes expected from mature IPs and none of the case studies contributed was submitted as a 'failure' case study. See the conclusion chapter for a discussion on what this current landscape of mature IPs implies.

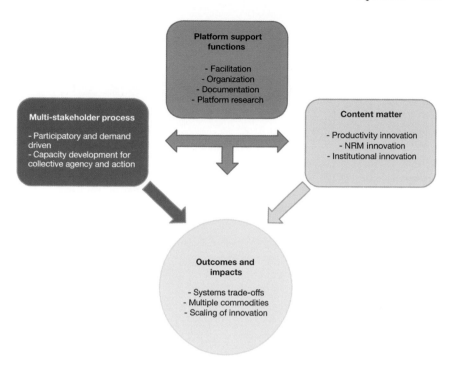

Figure 10.1 Relation between four key components of IP used to characterize the case studies

Multi-stakeholder processes help achieve IP outcome and impact

Participatory processes and demand-driven activities as starting point

The evidence from some of the case studies featured in this compilation supports previous findings that it is important for IPs to foster a participatory process that will lead to demand-driven activities, which in turn can contribute to achieving expected outcomes (Neef and Neubert, 2011; Cullen *et al.*, 2013; 2014; Leeuwis *et al.*, 2014). A recent evaluation of IPs operating in Ghana has shown that the IPs set up by the Volta2 project had led beneficiary farmers to identify the problems they would tackle together with other agricultural system stakeholders (Adane-Mariami *et al.*, 2013). The simple multi-stakeholder interactions within the Volta2 IPs in Ghana were enough to get farmers and other agricultural system stakeholders to take action in order to solve the farmers' problems. Adekunle and Fatunbi (2012) report similar dynamics.

The most notable example of this came from CIALCA in Burundi, DRC and Rwanda. The various national multi-stakeholder platforms emerged spontaneously to fulfil the mutual needs of farmers, government officials, the

Table 10.1 Case study categorization performed by the editors

Case study characteristics		Case study short names							
		CIALCA	NBDC	SysCom	MilkIT	We-RATE	NLA	Bubaare	Mukono–Wakiso
Multi-stakeholder processes	Participatory and demand driven	X			X				X
	Capacity development for collective agency and action	X					X	X	X
Content matter	Productivity innovation	X		X		X			X
	NRM innovation		X	X					
	Institutional innovation	X			X			X	
Platform support functions	Facilitation								X
	Organization							X	X
	Documentation						X		X
	Research on the platform								X
Outcomes and impact	Systems trade-offs		X						
	Multiple commodities					X		X	X
	Scaling of innovation					X			
	Failure								

Note: Editors scored each case study on a scale of 1 to 5. Cases that scored an average of 4 or higher are indicated in the table for the purpose of guiding the reader to cases that are particularly illustrative of the IP components listed in the first two columns.

private sector and other agricultural stakeholders around specific problems. This led to having different stakeholders in the platform, which was very useful for its facilitation and organization because several viewpoints, sources of knowledge and expertise were then available within the platform. As a further example, 25 participatory rural appraisals across the region helped identify entry points for platform activities. CIALCA IPs also organized field trials with pilot farmers and field visits for exchange of information between farmers and other agricultural stakeholders. The work to fight Banana Xanthomonas Wilt with Rwandan partners and with regional stakeholders in Rubavu was intrinsically a multi-stakeholder effort; that was how the platforms were constituted because all participants had an interest in solving the problem.

The cases of MilkIT in India and Mukono–Wakiso in Uganda also showed how initial multi-stakeholder meetings led to identifying gaps in resources and knowledge, which allowed the IP to co-opt other stakeholders to help solve the problem faced by farmers. Other similar examples are documented by Misiko (2014).

In contrast, other IPs featured in this compilation were still locked into a more top-down model of experts identifying the solution for the problem faced by farmers and providing their expertise through an IP, as also documented by Lynam and Blackie (1994). For example, NLA members in Nicaragua complained of not being able to influence the content of the agribusiness training delivered by the platform.

Capacity development for collective agency and action

As a space that promotes exchange of knowledge and learning between its members, IPs are a strong tool to develop the capacity of agricultural actors in working together to solve common problems (Ngwenya and Hagmann, 2011; Adekunle and Fatunbi, 2012). Thus, IPs that use their resources to build the capacities of their members are likely to establish their activities upon a solid base (World Bank, 2006; Misiko, 2014). The NLA case from Nicaragua provides the best example in this compilation of an IP investing in the capacity development of its members to give them more autonomy in their activities. Indeed, the whole objective of the NLA is to develop capacities in agribusiness management among farmers' cooperatives and individual farmer households. The NLA provided training on how to manage a farmers' group to representatives of the national farmers' groups who then snowballed the training within their own networks of local farmers' groups down to the individual farmers. This allowed the NLA to train representatives in 77 producers' organizations, reaching 19,347 households all over Nicaragua who were empowered to interact with market stakeholders. The NLA and its capacity development process thus achieved a tangible impact at scale through the large number of Nicaraguan farmers trained. Other examples of scaling capacity development through IPs are found in World Bank (2012).

While all the case studies featured in this compilation presented some capacity development activities, the cases of CIALCA and Bubaare provide other particularly good examples of how these activities within IPs can lead to impact. In particular, CIALCA's training of MSc and PhD students through its field research led to the development of strong linkages with important future decision makers within the AR4D system. These linkages facilitated subsequent CIALCA activities requiring strong support from government and national research centres. They also led to relevant policy changes that were informed by the IP's activities.

Appropriate content matter leading to platform impact

Productivity innovation to help farmers produce better

The diversity of actors involved within IPs makes them an ideal crucible of knowledge and innovative ideas to tackle complex agricultural development problems (Esparcia, 2014; Schut *et al.*, 2014). The mix between empirical scientific trials and indigenous stakeholder knowledge contributes to identifying the most appropriate techniques for productivity improvement. This state of cooperation to improve agricultural productivity has also been identified by Téno *et al.* (2013).

Among the case studies in this book, the tremendous impact reached through productivity innovation is best illustrated by WeRATE in West Kenya. The IP conducted farmer field trials for inoculant-fertilizer blend technology for a soybean variety. It disseminated its successful results to 37,000 farmer households and the potency of the technological innovation was confirmed by a 64 per cent adoption rate over four years of the N2Africa Project in which WeRATE participated. This case embodies the impact at scale that so many other IPs strive to achieve. It is partly thanks to its demonstrated superior technology, responsible for increases in farm productivity, which led to the widespread adoption and commercialization of some of the inputs or farming techniques developed by WeRATE partners.

The two Indian cases of SysCom and MilkIT show similar impact – though at a smaller scale – of innovative technologies suited to the local context of cotton and milk production, respectively. In particular, the innovative soil fertility management techniques developed and trialled by SysCom and its network of farmers were key to documenting the pros and cons of organic cotton production systems. Their productivity innovation trials thus provided the basis for informed decision making by farmers, between the trade-offs of conventional and organic systems. Likewise, CIALCA was particularly instrumental in advocating for the reintroduction of banana–coffee intercropping in Rwanda after its field trials demonstrated the superior quality of the coffee thus produced, with the caveat of a potentially heavier labour burden on women.

NRM innovation to strengthen the sustainability of agricultural systems

Previous studies have compiled the positive impact of IPs in tackling natural resources management challenges such as low soil fertility, low yields, erosion, deforestation and climate change (Darghouth *et al.*, 2008; Misiko *et al.*, 2013). One particularly telling example from this book is the NBDC project IPs in Ethiopia. The water bunds dug along steep slopes combined with new fodder production techniques shared by the project's researchers led to increased protection against soil erosion and raised awareness on linkages between production, marketing and NRM at community level. Although still small in scale, the local government has identified the NRM techniques trialled by the NBDC village IPs and wants to replicate this experience in other districts also affected by soil degradation and erosion.

Other good examples of NRM through IPs are provided by SysCom in India: they developed innovative methods for organic phosphorus fertilizer production for growing cotton. CIALCA also showed the example of integrating various crops on the hilly landscapes of Central Africa, which helped reduce soil erosion and led to more resilient cropping systems. Of the eight case studies featured in this compilation, only Bubaare and NLA have decided not to tackle environmental issues directly.

Institutional innovations provide the cement for replicability and marketability

Previous researches on IPs recognize their role in enhancing collaboration between stakeholders, developing social infrastructure, access to finance, certification, brokering land tenure arrangements, establishing public goods and markets, or simply leading to more relevant policy making (Paassen *et al.*, 2014). An independent study of the MilkIT project has shown how the IP had created the institutions necessary to get farmers planning their activities jointly with other agricultural system stakeholders, which led to increased marketing of their milk (Subedi *et al.*, 2014). This book also provides evidence of IPs emerging as a realistic route to creating and maintaining linkages across critical sector actors, who would usually act alone and with poor outcomes or impacts. Three of the case studies in this compilation provide very good examples of how institutional innovation contributes to development outcomes and impacts, as posited by Cortner *et al.* (1998).

For example, when considering the IPs' impact on national policy making, CIALCA provides a good example of how it adapted collaborative arrangements with its local partners according to the local institutional context. In Rwanda where government institutions are relatively strong and have a good presence across the territory, the government's research and extension system were key partners (Misiko, 2014). In Burundi and DRC, more complex arrangements linked government services and NGOs to allow innovations to reach potential end users. In all three countries, CIALCA also packaged its messages into suitable communication formats using local radio and video. Together with the targeted training of future policy makers already mentioned above, these institutional innovations led CIALCA to be an effective tool for policy influence in the three countries. At a more local level, the institutional collaboration fostered by the MilkIT IPs in Northern India have led to local government and agricultural support services, to coordinate their activities better to improve dairy production and marketing. Both IPs provide other good examples of how IPs can lead to more relevant policies for agricultural system actors (Cadilhon *et al.*, 2013).

The MilkIT project is an even more potent example of another institutional innovation fostered by an IP: linking smallholder farmers to markets. The farmers' cooperatives and milk collection centres that were fostered by the marketing IPs of the project have led to considerable market developments with more milk being produced for the market, new milk collectors and processors developing supply chains to the remote mountainous farm communities in Uttarakhand, and incomes increased throughout the dairy value chain. Further information on linking farmers to markets can be found in Birachi *et al.* (2013).

Finally, the institutional highlight of the Bubaare case from Uganda is the legal innovation of registering an IP as a new multi-purpose cooperative society (Makini *et al.*, 2013). This new legal framework for the IP mixes the benefits of farmers' cooperatives and multi-stakeholder groups. The new multi-commodity cooperative still allows farmers to group input purchases and output sales as in a regular farmers' cooperative but it also enables this grouping for several commodities. Furthermore, the loose network of suppliers, customers and other value-chain stakeholders who help commercialize the farmers' products around the cooperative are the product of the past multi-stakeholder activities of the IP. As a result of the legalization, the IP has been able to help its members gain the official standard certification from the Uganda National Bureau of Standards, allowing the farmers' various products to be sold to higher-end markets in Kampala city. Legalization has also enabled the farmers in the IP to do business with suppliers and customers on a larger scale while also improving their access to other services such as loans. As a result, the sorghum supply contract signed with Huntex Ltd has led to increasing sorghum supply from 500 kg per year by many individual farmers to reaching 2,000 kg per year supplied by just one entity: the cooperative. The farmers' organization as a cooperative with other value-chain actors within the IP remaining as advisors and business partners has increased the attraction power of IP membership for farmers: more than 1,000 individual farmers had joined the IP after just five years. During that period the membership of farmers' groups in the IP also increased from 32 groups to 1,121 groups, owing to its marketing services. This innovative legal framework for an IP will be particularly useful as a precedent for other countries sharing a common-law judicial tradition, in addition to other already existing legal statuses for multi-stakeholder commodity associations (Cadilhon and Dedieu, 2011).

Well-designed platform support functions leading to impact

The platform support functions include facilitation, organization, documentation and research on platforms. Besides purposefully connecting multistakeholder processes with the content matter, the platform support functions in themselves play an important role in allowing platforms to achieve impact at scale. One recent example of how platform support functions lead to impact comes from the Tanga Dairy Forum, where the organization of regular plenary and working committee meetings by the platform Secretariat have led to impacts in dairy development in this administrative region of Tanzania (Cadilhon, 2014). Likewise, the efforts of the Tanzania national Dairy Development Forum to facilitate communication and information sharing among members are fostering national dairy development (Kago *et al.*, 2015). And at the local level, the community-level IPs set up by the MilkIT sister project in Tanzania have also increased communication and information sharing among smallholder cattle herders, leading to improved access to, and better quality feeds (Pham Ngoc Diep *et al.*, 2015).

Of all the case studies featured in this compilation, the Mukono–Wakiso IP in Uganda is particularly noteworthy on all the aspects of platform support functions. Its facilitator from Makarere University has helped the platform members identify their priority entry points considering the needs of the farmers and to characterize the agricultural system combinations that would work within the set of entry points selected. By facilitating the platform's work with a systems perspective from the start, the platform agreed to work on an integrated system of crops, livestock and trees to help solve farmers' challenges. This has led directly to the IP's current work on multiple commodities. In terms of reporting, all Mukono–Wakiso IP events (not only formal platform meetings) result in a report stating the major decisions taken; they are shared with all members, and beyond, mainly using online repositories. This is essential to help the work of the chairman, facilitator and secretariat of the platform so as to follow up on tasks to be done to keep the activities going. The reports are useful for newcomers into the platform to catch up on previous activities and decisions. Finally, the Mukono–Wakiso platform is the object of research on multi-stakeholder processes by social scientists involved in the Humidtropics programme. This ongoing research is meant to help the platform in terms of its reflection on content, process and support functions in order to create impact in the future.

The Bubaare IP in Uganda has also demonstrated robust organization of its activities thanks to its new legal entity as a multi-commodity cooperative (see previous section). The NLA is another good example of documenting its agribusiness training process and monitoring progress in attaining expected outcomes: feedback from the trainees goes back up the training pyramid to the IP members who can then adjust their training materials and processes. The baseline data collected before the MilkIT project started in India and continuous data collection during the lifetime of the project have enabled MilkIT to gather strong evidence to present the concrete impacts highlighted in its case study.

Conclusion

This section has fleshed out examples of how multi-stakeholder process, content matter and support functions provided by IPs can lead to achieving impacts and outcomes in agricultural development. The data presented in Table 10.1 shows that most of the IPs featured in this compilation were particularly strong at fostering two out of the four components of the conceptual framework. Only Bubaare and Mukono–Wakiso IPs scored highly on all four of the IP processes. Nonetheless, even these two IPs have not yet shown achievements in all nine of the sub-components of successful IPs, nor have they managed to tackle all three sub-components of outcomes and impact, which would be expected from successful mature IPs.

Although it is not expected that IPs tackle all three of systems trade-offs, multiple commodities and scaling innovation outcomes at once, mature IPs should keep all three of these objectives in their sight. Therefore, we also have to conclude that the three components of IPs identified by the theoretical framework (process, content and platform support functions), posited to lead to platform outcomes and impact, are not sufficient factors of success. Rather, they are necessary factors for the IPs to be able to run and lead to some sustainable multi-stakeholder innovations. However, they do not provide guaranteed impact *at scale*. What would then be the additional factors to consider for IPs to achieve widespread impact on various commodities while still addressing the necessary trade-offs in complex multi-stakeholder agricultural systems? The concluding chapter of this book will provide some elements to start this discussion.

References

Adane-Mariami, Z., Cadilhon, J.J., Werthmann, C., 2013. Impact of innovation platforms on marketing relationships: the case of Volta Basin integrated crop-livestock value chains in Ghana. Paper presented at the Volta Basin Development Challenge Final Scientific Workshop, Ouagadougou, Burkina Faso, 17–19 September 2013.

Adekunle A.A., Fatunbi A.O., 2012. Approaches for setting-up multi-stakeholder platforms for agricultural research and development. *World Applied Sciences Journal* 16, 981–988.

Birachi, E., Rooyen, A.v., Some, H., Maute, F., Cadilhon, J., Adekunle, A., Swaans, K., 2013. Innovation platforms for agricultural value chain development. Innovation Platforms Practice Brief 6. ILRI, Nairobi, Kenya.

Cadilhon, J.-J., 2014. The Tanga Dairy Platform. Paper presented at the International Food and Agribusiness Management Association Forum, Cape Town, South Africa, 18 June 2014.

Cadilhon, J., Dedieu, M.-S., 2011. *Commodity Associations: A Widespread Tool for Marketing Chain Management*. Centre for Studies and Strategic Foresight Analysis no. 31. French Ministry of Agriculture, Paris, France.

Cadilhon, J., Birachi, E., Klerkx, L., Schut, M., 2013. Innovation platforms to shape national policy. Innovation Platforms Practice Brief 2. ILRI, Nairobi, Kenya.

Cortner, H.J., Wallace, M.G., Burke, S., Moote, M.A., 1998. Institutions matter: The need to address the institutional challenges of ecosystem management. *Landscape and Urban Planning* 40, 159–166.

Cullen, B., Tucker, J., Tui, S.H.-K., 2013. Power dynamics and representation in innovation platforms. Innovation Platforms Practice Brief 4. ILRI, Nairobi, Kenya.

Cullen, B., Tucker, J., Snyder, K., Lema, Z., Duncan, A., 2014. An analysis of power dynamics within innovation platforms for natural resource management. *Innovation and Development* 4, 259–275.

Darghouth, S., Ward, C., Gambarelli, G., Styger, C., Roux, J., 2008. Watershed management approaches, policies, and operations: lessons for scaling up. Water Sector Board Discussion Paper Series, Paper 11. World Bank, Washington, D.C.

Esparcia, J., 2014. Innovation and networks in rural areas: An analysis from European innovative projects. *Journal of Rural Studies* 34, 1–14.

Kago, K.M., Cadilhon, J.J., Maina, M., Omore, A., 2015. Influence of innovation platforms on information sharing and nurturing of smaller innovation platforms: A case study of the Tanzania Dairy Development Forum. Paper presented at the 29th International Conference of Agricultural Economists, Milan, Italy, 11 August 2015.

Leeuwis, C., Schut, M., Waters-Bayer, A., Mur, R., Atta-Krah, K., Douthwaite, B., 2014. Capacity to innovate from a system CGIAR research program perspective. Penang, Malaysia: CGIAR Research Program on Aquatic Agricultural Systems. Program Brief: AAS-2014–29, p. 5.

Lynam, J.K., Blackie, J.M., 1994. Building effective agricultural research capacity: The African challenge. In: Anderson, J.R. (ed.). *Agricultural Technology: Policy Issues for the International Community*. CABI, Wallingford, 106–134.

Makini, F.W., Kamau, G.M., Makelo, M.N., Adekunle, W., Mburathi, G.K., Misiko, M., Pali, P., Dixon, J., 2013. *Operational Field Guide for Developing and Managing Local Agricultural Innovation Platforms*. KARI and ACIAR, Nairobi, Kenya.

Misiko, M., 2014. Assessing and strengthening capacity for agricultural innovation platforms. Technical Report. A training workshop held at Hotel Des Mille Collines (Kempinski), Kigali, Rwanda, 15–19 September 2014. CIMMYT, Addis Ababa. Ethiopia.

Misiko, M., Mundy, P., Ericksen, P., 2013. Innovation platforms to support natural resource management. Innovation Platforms Practice Brief 11. ILRI, Nairobi, Kenya.

Neef, A., Neubert, D., 2011. Stakeholder participation in agricultural research projects: A conceptual framework for reflection and decision-making. *Agriculture and Human Values* 28, 179–194.

Ngwenya, H., Hagmann, J., 2011. Making innovation systems work in practice: Experiences in integrating innovation, social learning and knowledge in innovation platforms. *Knowledge Management for Development Journal* 7, 109–124.

Paassen, A.v., Klerkx, L., Adu-Acheampong, R., Adjei-Nsiah, S., Zannoue, E., 2014. Agricultural innovation platforms in West Africa. How does strategic institutional entrepreneurship unfold in different value chain contexts? *Outlook on Agriculture* 43, 193–200.

Pham Ngoc Diep, Cadilhon, J.J., Maass, B.L., 2015. Field testing a conceptual framework for innovation platform impact assessment: The case of MilkIT dairy platforms in Tanga region, Tanzania. *East African Agricultural and Forestry Journal* 81(1), 58–63.

Schut, M., van Paassen, A., Leeuwis, C., Klerkx, L., 2014. Towards dynamic research configurations: A framework for reflection on the contribution of research to policy and innovation processes. *Science and Public Policy* 41, 207–218.

Subedi, S., Cadilhon, J.J., Thanammal, R., Teufel, N., 2014. Impact evaluation of innovation platforms to increase dairy production: A case from Uttarakhand, northern India. Paper presented at the 8th International Conference of Asian Society of Agricultural Economists (ASAE) on Viability of Small Farmers in Asia 2014, Saver, Bangladesh, 15–17 August 2014.

Téno, G., Cadilhon, J.J., Somé, H., 2013. Impact of Volta2 innovation platforms on improvement and increase of crop and livestock production in four villages of Yatenga province, Northern Burkina Faso. Paper presented at the Volta Basin Development Challenge Final Scientific Workshop, Ouagadougou, Burkina Faso, 17–19 September 2013.

World Bank, 2006. *Enhancing Agricultural Innovation: How to Go Beyond the Strengthening of Research Systems.* World Bank, Washington, DC.

World Bank, 2012. *Agricultural Innovation Systems: An Investment Sourcebook.* World Bank, Washington, DC.

11 Are we there yet?

Some reflections on the state of
innovation platforms in agricultural
research for development

*Iddo Dror, Jean-Joseph Cadilhon and
Marc Schut*

Introduction

The previous chapter has linked together the underlying components of IPs with their performance. Using the eight case studies featured in this book as examples, we have identified how specific IPs have managed to make good use of their process, content and support functions in order to achieve impact at scale. However, it also highlighted that none of the platforms studied here had attained all three of the impacts expected from mature platforms: highlighting system trade-offs, generalizing activities to multiple commodities and reaching a large number of beneficiaries. This chapter presents the lessons learned by the case study authors for IPs to achieve impact. We also discuss areas of future research to identify the remaining factors that will lead IPs to deliver impact at scale.

In addition to the analysis based on the framework and matrices elaborated in the introduction and synthesis chapters, we also conducted interviews and facilitated exercises with all authors on what they considered to be the most important factors of success of IPs. This resulted in a common thread based on three complementing factors, namely *vision, enabling environment* and a *research for development orientation*. This concluding chapter will first provide a brief summary of each element, before proceeding to offer some final thoughts on the 'landscape' of mature IPs covered in this book, and some of the implications this holds for the future of IPs as a vehicle for agricultural development.

Success factors for IPs to achieve success

Vision

The first success factor that emerged was vision, or the fact that the IP should be clear about where it wants to go and how. To be successful, this vision should be embodied and encouraged by able leadership, which needs to be empowered and accountable for making sure that the IP focus of work 'emerges' from the commitment and common interest of participants rather than being 'established' through an external drive to tackle a problem.

In addition to able leadership, the group also identified skilful facilitation as another crucial element of the vision for IPs. The person facilitating the platform should be dedicated to this task and foster the participation of grass-roots actors from the bottom up, taking into account power dynamics. It is important for the facilitator to be physically present to participate regularly in platform activities as this helps foster trust between the platform members and between members and their facilitator.

Finally, the last component of vision is equity and transparency in the platform activities, whereby all actors in the platform are consulted in a similar way and all decisions taken have been discussed with the well-being of all actors in mind. Including equity and transparency in the platform vision helps strengthen the linkages between actors who are further motivated to participate.

Enabling environment

The second success factor of IPs involved in the case study competition was the enabling environment in which they thrived.

The first component of this enabling environment is the linkages with public policies. In some contexts, the coherence of the platform objectives with public policies has helped the platforms become essential to policy makers' engagement with grass-roots stakeholders for more relevant policy formulation and effective implementation. In other cases, IPs have supported the strengthening of public policies that were not appropriate to the local context by triggering the development of better policies. In line with coherence, some cases highlighted the importance of using already existing networks of stakeholders to foster innovations, rather than creating new platforms that duplicate work already being done in parallel multi-stakeholder groups.

The second component of an enabling environment for platforms is the willingness and capacity of members to participate in the innovation processes. This is achieved mainly through the skilful facilitation mentioned above and the search for right incentives, as discussed below. This involvement of all key stakeholders is particularly important for those who are likely to take action in order to reproduce successful innovations and disseminate them to other potential beneficiaries.

The third component of the enabling environment of IPs consists of the incentives that keep participants interested in contributing. These typically need to include short-term monetary incentives to attract and retain membership of smallholder farmers. However, a reachable mix of both short- and long-term expected benefits is more likely to sustain continuous motivation and participation from platform members.

Research for development orientation

The last success factor of the IPs reviewed in this compilation is the innovative science that the platform develops and trials. The application of applied science to solve real-life concrete problems and the participatory nature of the research trials conducted with platform stakeholders creates a meaningful link between science and practice.

To achieve this useful link, applying science on a joint and concrete problem faced by the platform members is the starting point. It is also useful to prioritize the research activities that are likely to generate quick results; this will foster the interest of platform participants and provide incentives for their further participation, as highlighted above. Participatory Action Research (PAR) is a useful approach to facilitating this type of embedded research for development.

The need for multifunctional IPs

Our synthesis demonstrated that none of the IPs featured in this compilation had attained all elements of impact at scale: systems trade-offs, application to multiple commodities and scaling of innovation (not to mention learning from failures). Therefore, we must ask why it is seemingly such an elusive task, and why platforms tend to gravitate towards a more narrow focus. Further research in this area, for example looking at the incentives and motivations of platform members, as well as their ability to manage multiple complex issues through a single entity would certainly be of interest in this context.

This section has fleshed out how the innovation process, innovation content, and support functions provided by IPs can lead to achieving impact in agricultural development. In the previous chapter, illustrative examples from the eight case studies featured in this compilation have demonstrated the links existing between these four elements of the theoretical framework, as proposed in the introductory chapter of this book. However, a closer look at the framework and its resulting impact matrix lead us to conclude that the three pillars identified by the theoretical framework (process, content and platform support functions), posited to lead to platform impact at scale, are prerequisite yet insufficient factors of success *at scale*. Yet, a definitive answer to what is the 'secret sauce' of IP success (if such even exists) will need to be the subject of further inquiries. Nevertheless, we can deduce the following conclusions.

Conclusion and final thoughts

As previously mentioned, we received no entries under the 'learning from failures' category. This in itself is a statement of sector immaturity, as it seems not to have embraced the approaches found in more mature sectors of owning up to failures and analysing to learn from both positive and negative lessons. A deeper look at the overall entries and cases published in this book further suggests that this is a trend that holds throughout. For example, we received only one entry on system trade-offs, and in the course of fleshing out the full case, its authors veered away from the core system trade-off elements to more generic productivity and process issues.

The two categories that had the bulk of the entries did not fully live up to what this process had targeted to showcase in terms of 'pure' entries in these categories. So for instance, multi-commodity cases were often a combination of crops, as opposed to the holistic crop–livestock–tree interactions that many researchers advocate. Likewise, the scaling cases were for the most part in the low thousands of direct outreach – not a small feat in some of the difficult environments where these platforms operate, but certainly not even a drop in the bucket when one thinks of the billions of farmers that large-scale initiatives aim to reach.

It is important to point out that useful elements emerged for each of these, even though they did not cut across the board – so while we see pockets of success, we still can't celebrate success across the board, or at a 'game-changer' scale. This then leads us to some of the questions we end up with, and which if/when answered, could provide a lot more insight into the suitability of IPs for specific work in a specific context, to inform investment decisions and facilitate more efficient and effective work in these areas. Some of these include:

• Why is the landscape the way it is? Our findings suggest that although most platforms are 'set up', as opposed to 'emerge', the scope of their focus areas still tends to be rather narrow, and somewhat in disconnect with the very holistic objectives promoted by those who set these IPs up. Could this be linked to short project cycles, the desire to show quick results, focus on short-term financial incentives, or a narrow focus of anchor projects, with no capacity to integrate broader and sustainable incentives? One of the key lessons from this exercise is that there is a need to avoid narrow processes, which requires that IPs become multifunctional by embracing multi-dimensional processes.

• Are IPs the most appropriate instrument to foster agriculture development? As was demonstrated through many of the cases, IPs can certainly lead to impact and can be an effective vehicle for agricultural development. However, it seems that insufficient attention has gone into examining whether the *solutions* developed by IPs (as opposed to the process), are scalable and replicable. Certainly, there is much to be said about the need for a much better availability of data on, and analysis of, the comparative return on investment (financial and otherwise) of IP work compared to a range of other intervention strategies. We've seen little evidence of a sense of urgency among researchers and practitioners alike to come up with a rigorous framework for measuring and reporting on this – but we feel that the lack of such an evidence-based approach is casting a shadow on much of the good work that is being showcased (including through anecdotal evidence such as most of the work presented in this book). How can this be measured? Similar exercises to understand more cases are critical to generate a matrix to guide any prudent investments in scaling approaches.

- Finally, when analysing the cases and framework findings, we emerged with a sense that IPs can potentially be a potent 'bridge' between the local ('small is beautiful') approaches that embody much of the participatory, demand-driven and community-led initiatives, and those global 'large scale impact' technology-driven initiatives.
- To be that bridge though, and to assume an integrative role for IPs alongside other approaches for inclusive agricultural development in the broader agricultural innovation system, the conceptual frameworks as well as the many implementation cases need to take a more balanced approach. They need to take into account local innovations but filter them (and the investment therein) through a lens of suitability for larger scale replication, and also factor in all direct and indirect costs to produce a more hard-nosed analysis of benefits per dollar invested.

Index

Page numbers in *italic* indicate tables; **bold** indicate figures, photographs and illustrations.